Advances in Fundamental and Applied Research on Spatial Audio

Edited by Brian F.G. Katz
and Piotr Majdak

Published in London, United Kingdom

IntechOpen

Supporting open minds since 2005

Advances in Fundamental and Applied Research on Spatial Audio
http://dx.doi.org/10.5772/intechopen.91556
Edited by Brian F.G. Katz and Piotr Majdak

Contributors

Amin Saremi, Wolfgang Klippel, Isaac Engel, Lorenzo Picinali, Brian F.G. Katz, David Poirier-Quinot, Peter Stitt, Martin S. Lawless, Zamir Ben-Hur, David Alon, Or Berebi, Ravish Mehra, Boaz Rafaely, Hans-Joachim Maempel, Michael Horn, Valentin Bauer, Amandine Pras, Leonard Menon, Dimitri Soudoplatoff, Katharina Pollack, Piotr Majdak, Wolfgang Kreuzer

Notice
Statements and opinions expressed in the chapters are these of the individual contributors and not necessarily those of the editors or publisher. No responsibility is accepted for the accuracy of information contained in the published chapters. The publisher assumes no responsibility for any damage or injury to persons or property arising out of the use of any materials, instructions, methods or ideas contained in the book.

First published in London, United Kingdom, 2022 by IntechOpen
IntechOpen is the global imprint of INTECHOPEN LIMITED, registered in England and Wales, registration number: 11086078, 5 Princes Gate Court, London, SW7 2QJ, United Kingdom
Printed in Croatia

British Library Cataloguing-in-Publication Data
A catalogue record for this book is available from the British Library

Additional hard and PDF copies can be obtained from orders@intechopen.com

Advances in Fundamental and Applied Research on Spatial Audio
Edited by Brian F.G. Katz and Piotr Majdak
p. cm.
Print ISBN 978-1-83969-005-1
Online ISBN 978-1-83969-006-8
eBook (PDF) ISBN 978-1-83969-007-5

We are IntechOpen,
the world's leading publisher of
Open Access books
Built by scientists, for scientists

6,000+
Open access books available

147,000+
International authors and editors

185M+
Downloads

Our authors are among the

156
Countries delivered to

Top 1%
most cited scientists

12.2%
Contributors from top 500 universities

Interested in publishing with us?
Contact book.department@intechopen.com

Numbers displayed above are based on latest data collected.
For more information visit www.intechopen.com

Meet the editors

Brian F.G. Katz is a research director at the French National Centre for Scientific Research, Sorbonne Université, Institut Jean Le Rond d'Alembert, in the group Lutheries - Acoustics - Music. His fields of interest include spatial 3D audio rendering and perception, room acoustics, HCI, and virtual reality. With a background in physics and philosophy, he obtained his Ph.D. in Acoustics from Penn State in 1998 and his HDR in Engineering Sciences from UMPC in 2011. Before joining CNRS, he worked for various acoustic consulting firms, including Artec Consultants Inc., ARUP & Partners, and Kahle Acoustics. He has also worked at Laboratoire d'Acoustique Musical (UPMC), IRCAM, and LIMSI-CNRS.

Piotr Majdak studied electrical and audio engineering at the University of Technology and the University of Music and Performing Arts, both in Graz, Austria. He is the deputy director of the Acoustics Research Institute (ARI) of the Austrian Academy of Sciences (OeAW) where he is a member of the psychoacoustics and experimental audiology group. He received his Ph.D. in Psychoacoustics and Signal Processing and is a professor of acoustics and audio engineering. He regularly teaches at various universities. His main interest is the understanding of spatial hearing in humans linked to its various applications in spatial-audio reproduction systems.

Contents

Preface

Spatial audio is a dynamic and rapidly evolving field, the consequence of being closely linked to advances in computer technology and digital signal processing. The democratisation of virtual reality hardware available as consumer devices has moved the field towards applications and further away from traditional laboratory research. This book, Advances in Fundamental and Applied Research on Spatial Audio, includes eight peer-reviewed chapters on this exciting area of research. The chapters are organised into three sections: "Acoustic Methodology," "Perception," and "Applications."

The first section addresses advances related to both loudspeaker and headphone presentation in the context of spatial audio. For loudspeaker reproduction, the first chapter in this section discusses requirements for 3D audio from the perspectives of physical modelling and output-based measurements, considering the special requirements for sound-field control and including new metrics that simplify the interpretation of loudspeaker properties at individual points, sound zones, and across the entire sound field. For headphone reproduction, the second chapter provides an overview of state-of-the-art methods for capturing personalised head-related transfer functions (HRTFs).

The second section addresses binaural spatial perception, considering its quantification, multimodal interactions, and the limitations of reverberation perception. The first chapter in this section presents a meta-analysis of raw data from several studies and discusses HRTF performance evaluation methods and metrics, highlighting issues in both the design of evaluation protocols and the selection of metrics for better comparisons between studies. The second chapter examines the influences of the presence and characteristics of acoustic and visual environments on the perceived distance and room size, employing an extra-aural headset and a semi-panoramic stereoscopic projection. The third chapter provides an overview of recent research on reverberation perception in a binaural surround-sound context, with a focus on how to enable more efficient reproduction of realistic sound scenes. Special emphasis is given to Ambisonics-based techniques and the effect of spatial resolution on the perceptual quality of binaural reproduction.

The third and final section focuses on applied research and includes three chapters. The first chapter presents a method for improved rendering of Ambisonic sound fields incorporating ear orientation under the term Bilateral Ambisonics, an Ambisonic representation of the sound field formulated at both ears. The second chapter examines, in the context of modern automotive acoustic and audio environments, the fundamental and practical aspects of acoustic echo cancellation, noise reduction, reverberation reduction, and beamforming signal processing methods, with the aim of spatially enhancing signals and creating listening zones in cars. The final chapter addresses several case studies conducted in various settings, comparing the experiences of orchestra conductors and instrumentalists monitoring their performances with binaural and stereo headphone-based sound reproduction.

We gratefully acknowledge the support of the SONICOM project (www.sonicom.eu), funded under the European Union's Horizon 2020 research and innovation program under grant agreement No. 101017743, and the Aural Assessment By means of Binaural Algorithms (AABBA) working group in the preparation of this anthology.

Brian F.G. Katz
Institut Jean Le Rond d'Alembert,
Sorbonne Université,
CNRS,
Paris, France

Piotr Majdak
Acoustics Research Institute,
Austrian Academy of Sciences,
Vienna, Austria

Section 1

Acoustic Methodology

Chapter 1

Modeling and Testing of Loudspeakers Used in Sound-Field Control

Wolfgang Klippel

Abstract

This chapter describes the physical modeling and output-based measurement of loudspeakers, essential hardware components in sound-field control. A gray box model represents linear, time-variant, nonlinear, and non-deterministic signal distortions. Each distortion component requires a particular measurement technique that includes test stimulus generation, sound pressure measurement at selected points in 3D space, and signal analysis for generating meaningful metrics. Near-field scanning measures all signal components at a large signal-to-noise ratio with minor errors caused by loudspeaker positioning, air temperature, room reflections, and ambient noise. Holographic postprocessing based on a spherical wave expansion separates the direct sound from room reflections to assess the linear output and signal distortion. New metrics are presented that simplify the interpretation of the loudspeaker properties at single points, sound zones, and over the entire sound-field.

Keywords: loudspeaker directivity, near-field scanning, signal distortion, nonlinear loudspeaker modeling, sound-field control, spatial sound application

1. Introduction

Loudspeakers play an essential role in spatial sound applications, such as conventional multi-channel sound reproduction, beam steering [1], wave-field reconstruction [2], higher-order ambisonics [3], immersive audio [4], and multi-zone contrast control [5]. Those techniques require many loudspeakers arranged in linear, planar, circular, and spherical arrays [6] to satisfy the spatial sampling theorem at higher frequencies and provide desired directivity, sufficient sound power output, and audio quality. Cost, size, weight, and energy consumption are critical factors limiting the practical application.

Sound-field control techniques can use model-based or data-based methods to calculate the individual driving signals for the loudspeakers. Both approaches prefer an idealized loudspeaker model, usually assuming a linear, time-invariant transfer behavior and omnidirectional radiation while ignoring undesired properties (e.g., distortion) and physical limitations of the loudspeaker.

Loudspeakers are not always omnidirectional, especially at high frequencies. Various theories [7–9] consider and exploit the loudspeaker directivity in sound-field control. There are exciting opportunities for loudspeaker arrays exploiting a higher-order spherical wave model used in reverberant rooms [10].

Standard characteristics describe the loudspeaker directivity in the far-field [11]. Still, this information is less relevant in applications for home, automotive, or public address systems where either the radiating surface is large (e.g., arrays, flat panel) or the distance to the listener is small. Choi et al. [12] showed that active control could cope with those conditions if the near-field properties of the loudspeaker are considered.

Xiaohui et al. [13] showed that loudspeaker nonlinearities degrade the performance of spatial sound control, as nonlinear distortions limit the acoustic contrast between "bright" and "dark" sound zones. Cobianchi et al. [14] proposed a method for measuring the directivity of the nonlinear distortion in the far-field by using sinusoidal and multi-tone stimuli. Such tests performed in the near and far-field generate a significant test effort and a high amount of data that can be difficult to interpret.

Olsen and Møller [15] showed that typical ambient temperature variations in automotive applications change the loudspeaker properties in ways that compromise the sound zone performance significantly. Production variability, heating of the voice coil, fatigue, and aging of the suspension and other soft parts (cone) can change the loudspeaker properties over time and degrade the performance in a non-adaptive control solution.

This chapter presents models and measurement techniques to assess the loudspeaker transfer behavior from the input to the sound pressure at any point in the sound-field. The objective is to generate comprehensive information for selecting loudspeakers for spatial sound applications, simulating the performance, including room interaction, and maintaining sound quality over product life.

Such measurements are intended to provide meaningful characteristics that describe the sound pressure at a local point, over a listening zone, or in all directions, simplifying loudspeaker diagnostics.

2. General loudspeaker modeling

A single loudspeaker system used in spatial audio applications can be modeled by a multiple-input-multiple-output system (MIMO), as shown in **Figure 1**.

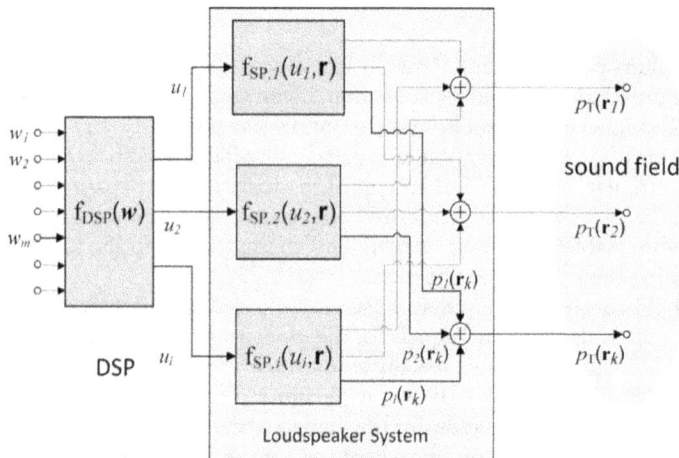

Figure 1.
Modeling a loudspeaker system with multiple channels in spatial sound applications.

The loudspeaker input signals

$$u_i = \mathrm{f}_{\mathrm{DSP}}(w_1, w_2, \dots, w_{N_{\mathrm{DSP}}}) \quad i = 1, \dots, N_u \tag{1}$$

are generated by sound-field control or other DSP algorithms $\mathrm{f}_{\mathrm{DSP}}$ applied to audio signals w_{m}. The input signal u_i can be an analog voltage at the loudspeaker terminals or a digital data stream using other electrical, optical, or wireless transmission means. For each input signal u_i, the loudspeaker system uses at least one electro-acoustical transducer (woofer, tweeter, full-band driver) that generates a sound pressure $p_i(\mathbf{r})$ at an evaluation point $\mathbf{r}$ under free-field condition. In modern loudspeaker systems, the transduction block $\mathrm{f}_{\mathrm{SP},i}(u_i,\mathbf{r})$ also performs amplification, equalization, active speaker protection against mechanical and thermal overload [16], and adaptive nonlinear control to cancel undesired signal distortion [17]. The total sound pressure output $p_T(\mathbf{r})$ is a linear superposition of the contributions $p_i(\mathbf{r})$ from all transduction blocks described as

$$p_T(\mathbf{r}) = \sum_{i=1}^{N_u} p_i(\mathbf{r}) = \sum_{i=1}^{N_u} \mathrm{f}_{\mathrm{SP},i}(u_i, \mathbf{r}) \tag{2}$$

while assuming a negligible coupling between the loudspeaker channels in the electrical, mechanical, or acoustical domain. This assumption is valid for transducers radiating sound independently into the free-field but not for multiple transducers mounted in one enclosure and working on the same air volume.

The function $\mathrm{f}_{\mathrm{SP},i}(u_i, \mathbf{r})$ describes the nonlinear and time-variant relationship between input u_i and output signal $p_i(\mathbf{r})$.

The following chapter describes a single loudspeaker channel's modeling, measurement, and quality assessment while omitting the subscript i in the input voltage u ($u_j = 0$ for $j \neq i$) and the sound pressure output $p(\mathbf{r})$.

Figure 2 shows a gray box model representing the nonlinear, time-variant function $\mathrm{f}_{\mathrm{SP}}(u, \mathbf{r})$ under free-field conditions. At small input signal amplitudes, the linear spatial transfer function $H_L(f,\mathbf{r})$ describes the loudspeaker behavior, assuming that other signal distortions are negligible. Still, additional noise $n(\mathbf{r})$ generated by electronics or external sources can corrupt the sound pressure output.

The time-variant transfer function $H_V(f,t)$ represents reversible and nonreversible changes in the loudspeaker properties caused by the stimulus, climate [15], heating [18], aging, fatigue [19], and other external influences. The function $H_V(f,t)$ is independent of the evaluation point $\mathbf{r}$ because the dominant time-variant processes are in the electrical and mechanical domains. For example, the voice coil resistance [18], the natural frequencies, and loss factors of the modal vibrations [20] affect the sound-field in the same way. Variations of the mode shape, box geometry, and other boundaries can change the loudspeaker directivity but are

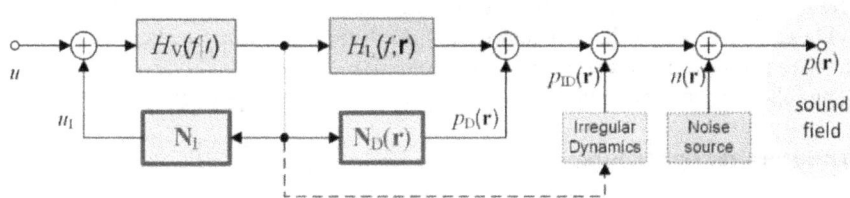

Figure 2.
Gray box model of a single loudspeaker channel describing the relationship between the input signal u and sound pressure output p(r) at an evaluation point r in the free-field.

neglected in the modeling. The $H_V(f,t)$ variation can be monitored by endurance, environmental or accelerated-life testing defined in various loudspeaker standards [11, 21].

Nonlinear subsystem N_I and N_D generate harmonics and intermodulation distortions at higher amplitudes. The first nonlinear system N_I in the feedback loop in **Figure 2** represents the dominant nonlinearities [22] in the transduction and the mechanical suspension such as force factor, voice coil inductance, and stiffness of a moving coil speaker [23]. A network with lumped parameters models the nonlinear dynamics by generating equivalent input distortion u_I added to the input signal u and transferred via the linear transfer path to any point **r** in the sound-field [11].

The second nonlinear subsystem $N_D(\mathbf{r})$ in **Figure 2** represents nonlinearities in the cone, diaphragm, surround, horn, port, and other acoustic elements and generates distributed distortion $p_D(\mathbf{r})$. The distributed distortion $p_D(\mathbf{r})$ depends on the point **r** and cannot be represented by equivalent input distortion.

The nonlinear distortions u_I and $p_D(\mathbf{r})$ are considered in loudspeaker design because they affect the maximum output, audio quality, size, cost, and reliability. Finally, the distortions accepted as regular properties give the best performance-cost ratio for the end-user.

Imperfections in the design, manufacturing problems, overload, and other malfunction ("rub&buzz") generate irregular dynamics perceived as abnormal distortion $p_{ID}(\mathbf{r})$ that is partly not deterministic and not predictable.

3. Acoustical loudspeaker measurements

The free model parameters and other signal-dependent characteristics introduced in the gray box model presented in Section 2 can be identified by acoustic measurements.

The sound pressure can be modeled as a superposition of desired and undesired signal components in the time domain as

$$p(t,\mathbf{r}) = p_L(t,\mathbf{r}) + p_V(t,\mathbf{r}) + p_N(t,\mathbf{r}) + p_{ID}(t,\mathbf{r}) + n(t,\mathbf{r}) \qquad (3)$$

and in the frequency domain as a corresponding Fourier spectrum:

$$\begin{aligned} P(f,\mathbf{r}) &= F\{p(t,\mathbf{r})\} \\ &= P_L(f,\mathbf{r}) + P_V(f,\mathbf{r}) + P_N(f,\mathbf{r}) + P_{ID}(f,\mathbf{r}) + N(f,\mathbf{r}) \end{aligned} \qquad (4)$$

The component p_L represents the desired linear output separated from signal distortion components p_V, p_N, p_{ID}, and n corresponding to the time-variant properties, regular loudspeaker nonlinearities, and abnormal distortion generated by irregular vibration and measurement noise, respectively.

New output-based measurement techniques compliant with IEC 60268–21 [11] provide accurate data with sufficient spatial resolution in a non-anechoic environment with minimum test effort (time, equipment).

The following sections will discuss those signal components in greater detail.

3.1 Loudspeaker positioning

The positioning of the loudspeaker in the 3D space is clearly defined by IEC 60268–21 [11] using a spherical coordinate system using the polar angle θ, azimuthal angle ϕ, and distance r. The origin **O** is placed at a convenient reference point $\mathbf{r}_{ref}$, usually on the radiator's surface, grill, or enclosure, close to the supposed acoustical center. A reference axis $\mathbf{n}_{ref}$ is orthogonal to the radiator's surface, and the orientation vector $\mathbf{o}_{ref}$ usually points upwards in a vertical direction.

3.2 Test environment

To ensure the reproducibility of the test result, it is common practice to measure loudspeakers under free-field conditions using a full-space (4π) or half-space (2π) environment. A half-space anechoic room with a solid ground floor is convenient for moving large and heavy loudspeaker systems and measuring loudspeakers mounted in or placed at a short distance from walls. The IEC standard [11] defines various methods of testing and postprocessing to generate simulated free-field conditions in a non-anechoic environment.

3.3 Far-field measurement

The traditional way to assess the loudspeaker directivity is the measurement of the spatial transfer function $H_L(f, r_D, \theta, \phi)$ between the input u and the sound pressure output $p(r_D, \theta, \phi)$ under far-field condition [11]. The distance r_D between the loudspeaker and microphone should be much larger than the size of the speaker and acoustic wavelength. The $1/r$ law valid in the far-field allows extrapolating the complex transfer function to other distances r as

$$H_L(f, r, \theta, \phi) = H_L(f, r_D, \theta, \phi) \frac{r_D}{r} e^{-jk(r-r_D)} \tag{5}$$

using the wavenumber $k = 2\pi f/c_0$ and the speed of sound c_0. Large loudspeakers such as loudspeaker arrays, soundbars, flat-panel speakers, and horn loudspeakers require a large measurement distance r_D and a sizeable anechoic room with good air conditioning to keep the variance of the temperature field sufficiently small.

The choice of measured directions determines the angular resolution of the directional gain [11], the accuracy of coverage angle [11], and other derived far-field characteristics. 2-degree angular resolution, needed for some professional loudspeakers, requires about 16,000 measurement points. Rotating a large and heavy loudspeaker over all combinations of the two angles requires robust and accurate robotics with speed ramps to accelerate and deaccelerate the mass. A microphone array speeds up the test by simultaneously measuring the sound pressure at multiple points without moving the loudspeaker.

Common far-field measurements usually provide no information about the accuracy of the measured data. They cannot indicate errors related to the positioning of loudspeakers or microphones, insufficient sampling of complex directivity patterns, or acoustical disturbances due to wind, air temperature, static sound pressure, or ambient noise [15].

Minor positioning errors and normal variation of the speed of sound, which is usually not critical for the amplitude response, can cause significant errors in the phase response and degrade the performance of 3D sound applications. For example, a deviation of the room temperature by 2 Kelvin during the test changes the speed of sound by 1.2 m/s and the acoustic propagation time by 50 μs at a measurement distance $r = 5$ m, which is required to ensure far-field condition for large loudspeakers. This time delay corresponds to a positioning error of 17 mm and generates a phase error of 36 degrees at 2 kHz, increasing linearly with frequency and reaching 180 degrees at 10 kHz.

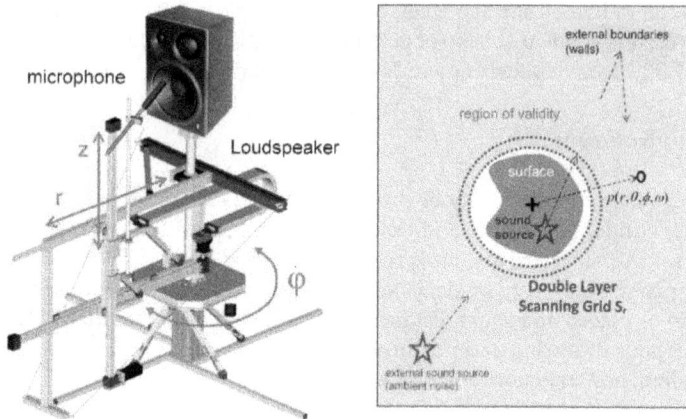

Figure 3.
Nearfield measurement by placing the loudspeaker at a fixed position and moving a microphone with robotics over the scanning grid close to the speaker surface.

3.4 Near-field measurement

The IEC standard 60268-21 [11] recommends measurements in the near-field, which overcome the restrictions and problems faced in the far-field. However, the $1/r$ law in Eq. (5) is not applicable, and a holographic measurement technique that scans the sound pressure and fits a spherical wave model to measured data is required.

Figure 3 shows a scanning system used for measuring the sound pressure generated by a loudspeaker placed at a fixed position on a post. The microphone moves in three axes in cylindrical coordinates (r,φ,z) to multiple test points $r_k \in S_r$ distributed on a double layer grid S_r close to the speaker's surface [24]. Moving a lighter microphone instead of rotating the heavier loudspeaker simplifies the robotics, allows faster speed ramps, and reduces the positioning error. Those opportunities make it possible to generate redundancy in the collected data and check the measurement's accuracy.

The scanning points are distributed on two concentric layers, as shown in **Figure 3**, to measure the local derivative of the sound pressure like a sound intensity probe. That is the basis for separating the outgoing wave comprising direct sound radiated by the loudspeaker (e.g., diaphragm) from the incoming wave generated by reflections on the positioning arm of the robotics, ground floor, and room walls. The close distance to the sound source increases the direct sound, which increases the signal-to-noise ratio (SNR) by more than 20 dB and significantly reduces the phase error caused by varying air properties in far-field measurements.

4. Spatial transfer function

The spatial transfer function $H_L(f,\mathbf{r})$ describes the linear relationship between input spectrum $U(f)$ and sound pressure spectrum $P_L(f,\mathbf{r})$ generated by the loudspeaker at any point $\mathbf{r}$ under the free-field condition as a spherical wave expansion in Eq. (6) using general solutions $\mathbf{B}_{out}(f, \mathbf{r})$ of the Helmholtz equation weighted by complex coefficients in vector $\mathbf{C}_L(f)$ [25]:

$$H_L(f,\mathbf{r}) = \frac{P_L(f,r,\theta,\phi)}{U(f)} = \mathbf{C}_L(f)\mathbf{B}_{OUT}(f,\mathbf{r})$$

$$= \sum_{n=0}^{N}\sum_{m=-n}^{n} c_{n,m}^L(f)h_n^{(2)}(kr)Y_n^m(\theta,\phi)$$

(6)

The spherical coordinates allow a separation of angular dependency using the spherical harmonics $Y_n^m(\theta,\phi)$ from the radial dependency using the Hankel function of the second kind $h_n^{(2)}(kr)$. The spherical harmonics have orthonormal properties representing a monopole ($n = 0$), dipoles ($n = 1$), quadrupoles ($n = 2$), and more complex sources with increasing order n.

Figure 4 illustrates the expansion for a woofer operated in a sealed enclosure at 200 Hz. The measured directivity pattern is presented as a target on the lower left-hand side and compared with the wave model for rising maximum order N. The expansion can be truncated at $N = 3$ because 16 coefficients weighting the spherical harmonics provide sufficient accuracy. Higher-order terms can be ignored at 200 Hz because they are 50 dB below the total sound power. The contribution of the higher-order terms rises with frequency and is required to explain the directivity pattern at 1 kHz, as shown in the upper diagram on the right-hand side.

The Hankel function $h_n^{(2)}(kr)$ in Eq. (6) models the decay of the sound pressure with rising distance r from expansion point $\mathbf{r}_e$ of the spherical wave expansion. In the near-field for $r < r_{far}$, the $1/r$ law is not valid anymore because sound pressure and particle velocity are not in phase, generating an increase in the apparent power at lower distances [24]. In the far-field $r > r_{far}$, the sound pressure decreases inversely with the rising distance r giving 6 dB less output for doubling the distance. Thus, the apparent sound power radiated from the loudspeaker is constant and corresponds to the real power.

Figure 5 shows the power $\Pi_n(r)$ contributed by spherical waves of order n to the total apparent power $\Pi_a(r)$. Only the order $n = 0$ (monopole) generates a constant power output for all distances while the steepness of the power curve $\Pi_n(r)$ in the near-field increases with the order n of the waves.

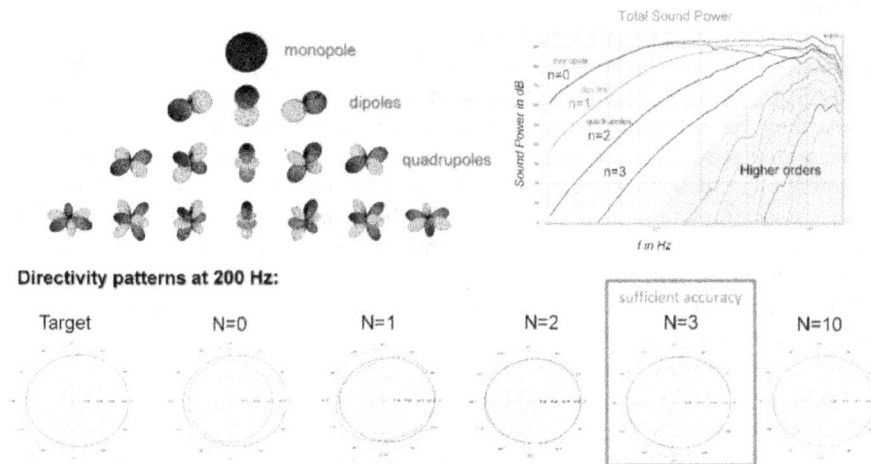

Figure 4.
Modeling the total sound power frequency response (upper right) and directivity pattern at 200 Hz (below) of a loudspeaker by spherical wave model (upper left).

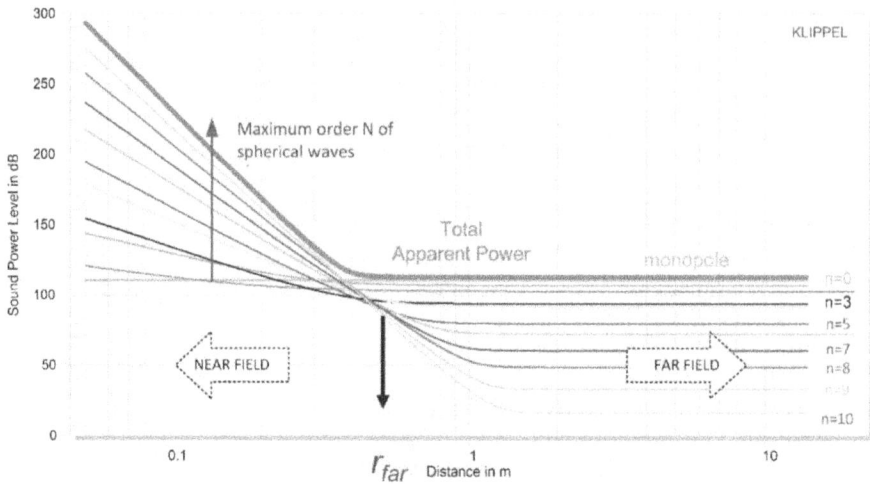

Figure 5.
Total apparent sound power $\Pi_a(r)$ (thick line) generated by a loudspeaker versus radial distance r and the contribution $\Pi_n(r)$ of the spherical waves of order n (thin lines).

4.1 Parameters of the linear model

The optimum coefficients $\mathbf{C}_L(f)$ in the spherical wave model in Eq. (6) can be calculated by minimizing the mean squared error between the response $H_L'(f,\mathbf{r}_k)$ measured at scanning points $\mathbf{r}_k \in S_r$ and the modeled responses as

$$\mathbf{C}_L(f) = \underset{\mathbf{C}}{\arg\mathrm{MIN}} \sum_{k=1}^{K_r} \left| H_L'(f,\mathbf{r}_k) - \mathbf{C}(f)\mathbf{B}_{\mathrm{OUT}}(f,\mathbf{r}_k) \right|^2 \qquad (7)$$

Normalizing the mean squared error in Eq. (7) with the total output power gives a valuable criterion e for checking the measurement's spherical wave expansion accuracy [24].

Figure 6 shows the normalized fitting error e in the wave expansion with rising total order N. A single monopole expansion ($N = 0$) already gives an error reduction of 10 dB at 100 Hz. Considering the monopole and the three dipoles ($N = 1$) can reduce the error to minus 20 dB at 100 Hz, which means the model can explain 99% of the output power. A wave expansion of order $N = 5$ requiring at least 36 measurement points describes the sound output of the woofer channel below 1 kHz with sufficient accuracy ($e < 1\%$). The increase of the fitting error at higher frequencies indicates that higher-order terms are required in the expansion to model the directivity at higher frequencies.

This example shows that the loudspeaker properties determine the maximum order N of the expansion, the number of measurement points K_r required to identify the coefficients $\mathbf{C}_i(f)$, and the total scanning time.

For acoustic, esthetic, or technical reasons, most loudspeakers have a natural symmetry in the diaphragm's shape, the cone placement on the front side of the cabinet, and the enclosure's geometry. Symmetry factors [24] calculated from identified coefficients $\mathbf{C}_L(f)$ during the scanning process reveal the loudspeaker's left/right or top/bottom single-plane, dual-plane or rotational symmetry. This information can be used to align the loudspeaker position and orientation with spherical harmonics to reduce the number of measurement points required to fit the wave

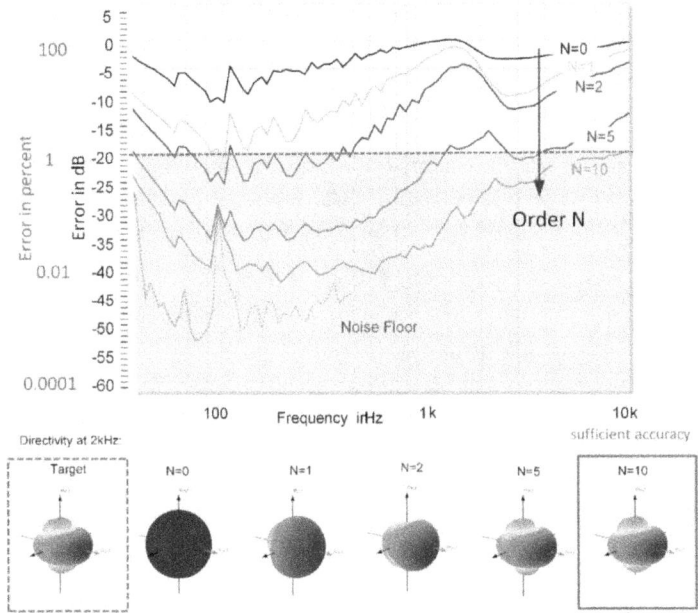

Figure 6.
Normalized fitting error e versus frequency f of the spherical wave expansion truncated at maximum order N (above) and corresponding identified directivity pattern shown as a balloon-plot for the corresponding order N compared with the measured target response (left-hand side below).

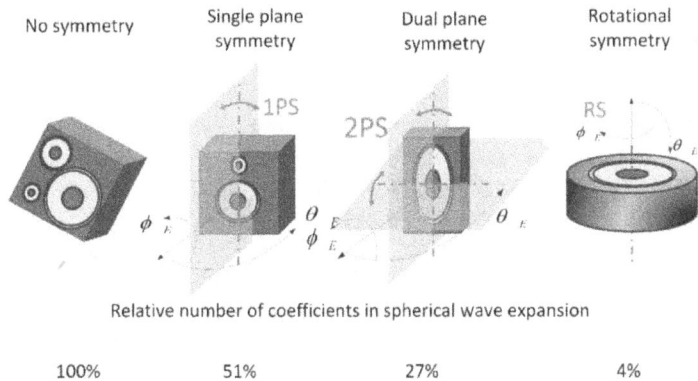

Figure 7.
Exploiting symmetry in the loudspeaker geometry to reduce the number of measurement points required for the spherical wave expansion.

expansion. As illustrated in **Figure 7**, considering the rotational symmetry can reduce the number of measurement points to 4%, significantly speeding up the scanning process.

4.2 Simulated free-field condition

The measurement of the spatial transfer function requires free-field conditions or at least simulated free-field conditions as defined in IEC standard 60268–21 [11]. The absorption of the lined walls in "anechoic" rooms is usually imperfect at low frequencies where the wavelength of the standing waves exceeds the thickness of the lining. Gating the sound pressure signal and windowing of the impulse response provides good results at higher frequencies but degrade the frequency resolution at low frequencies.

The wave separation technique based on near-field scanning on two surfaces [25] can be used to separate the direct sound from the room reflections at low and middle frequencies and complements the windowing technique at higher frequencies. The measured transfer function $H'_L (f, \mathbf{r}_k)$ with $\mathbf{r}_k \in S_r$ corrupted by room reflections can be modeled by a spherical wave expansion [26]

$$H'_L(f, \mathbf{r}_k) = \mathbf{C}(f)\mathbf{B}(f, \mathbf{r}_k) = \mathbf{C}_L(f)\mathbf{B}_{OUT}(f, \mathbf{r}_k) + \mathbf{C}_{SR}(f)\mathbf{B}_{SR}(f, \mathbf{r}_k)$$

$$= \sum_{n=0}^{N} \sum_{m=-n}^{n} \left(c_{n,m}^{out} h_n^{(2)}(kr) + c_{n,m}^{SR} J_n(kr) \right) Y_n^m(\theta, \varphi) \tag{8}$$

considering outgoing wave $\mathbf{B}_{OUT}(f, \mathbf{r}_k)$ radiated by the loudspeaker as used in Eq. (6) and reflected waves $\mathbf{B}_{SR}(f, \mathbf{r}_k)$ represented by Bessel functions of the first kind $J_n(kr)$. The optimal coefficients $\mathbf{C}_L$ and $\mathbf{C}_{SR}$ minimizing the mean squared error between measured and modeled response can be estimated by

$$\mathbf{C}(f) = [\mathbf{C}_L(f) \quad \mathbf{C}_{SR}(f)]$$

$$= \arg\MIN_{\mathbf{C}} \sum_{k=1}^{K_r} |H'_L(f, \mathbf{r}_k) - \mathbf{C}(f)\mathbf{B}(f, \mathbf{r}_k)|^2 \tag{9}$$

The coefficients $\mathbf{C}_{SR}(f)$ provide the SPL response of the sound reflections shown as a dashed curve in **Figure 8** that corrupts the measurement and causes a significant error below 1 kHz in the measured SPL response (thin green solid line). The $\mathbf{C}_L(f)$ represents the SPL direct sound (thick blue solid line) measured under simulated free-field conditions.

4.3 Interpretation of the spatial transfer function

The interpretation of the spatial transfer function $H_L(f, \mathbf{r})$ can be simplified by calculating the SPL frequency response at point $\mathbf{r}$ in decibel as

$$L_{SP}(f, \mathbf{r}) = 20\lg\left(\frac{|H_L(f, \mathbf{r})|\tilde{u}}{p_{ref}}\right) dB \tag{10}$$

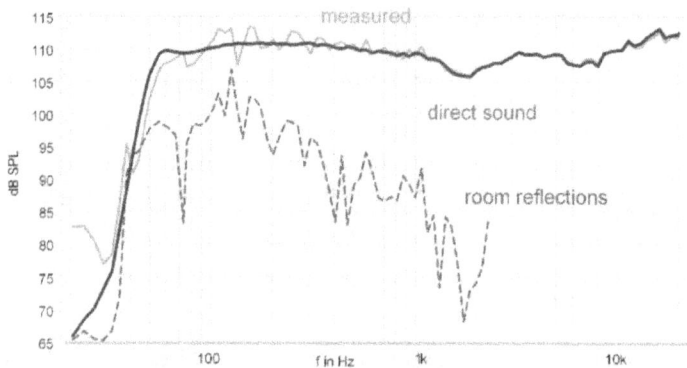

Figure 8.
Generating simulated free-field conditions at low frequencies by separating direct sound (solid line) from the room reflections (dashed line) in the measured SPL frequency response (thin line).

using a fixed RMS value $\tilde{u}$ of the input signal $u(t)$ and the reference sound pressure $p_{ref} = 20\mu Pa$. The SPL frequency response displayed in 2D or 3D plots (polar, balloon, contour) shows the directional dependency versus angles θ and ϕ in the far-field $r > r_{far}$ as shown in **Figure 9** and the local dependence versus Cartesians coordinates x,y,z in the near and far-field in **Figure 10**.

The phase response at point **r** calculated as

$$\varphi(f,\mathbf{r}) = \arg(H_L(f,\mathbf{r}))$$
$$= \varphi_M(f,\mathbf{r}) + \varphi_A(f,\mathbf{r}) - 2\pi f\tau(\mathbf{r}) \tag{11}$$

provides essential information for combining multiple loudspeaker channels in systems and arrays and applying DSP processing to control the sound-field. The total phase response $\varphi(f,\mathbf{r})$ can be decomposed into three parts: The minimal phase $\varphi_M(f, \mathbf{r})$ corresponds to the amplitude response $|H_L(f, \mathbf{r})|$ via the Hilbert Transform. The all-pass phase $\varphi_M(f,\mathbf{r})$ reveals the polarity and other loudspeaker properties. A critical part is a total time delay

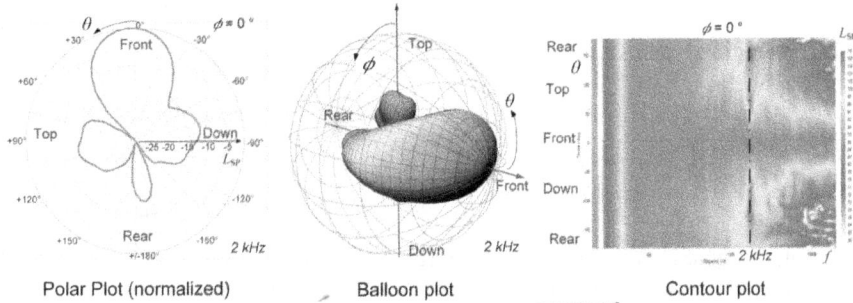

Figure 9.
Visualization styles for the far-field directional SPL response $L_{SP}(f,r,\theta, \phi)$ in spherical coordinates.

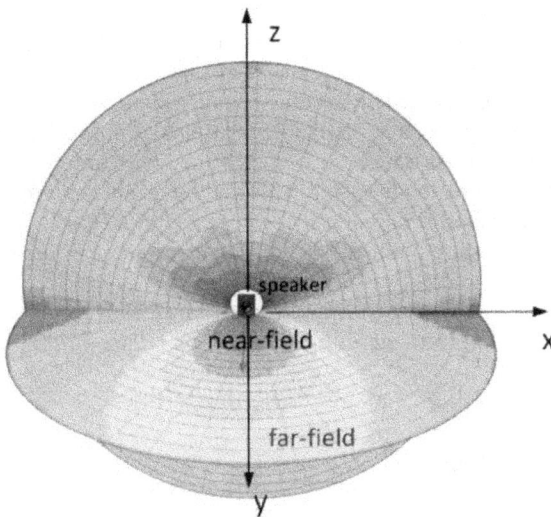

Figure 10.
Visualization of the SPL of the direct sound-field $L_{SP}(f,x,y,z)$ generated by a loudspeaker at 2 kHz outside the scanning surface.

$$\tau(\mathbf{r}) = \tau_{\text{DSP}} + \frac{|\mathbf{r} - \mathbf{r}_e|}{c_0(T_A(\mathbf{r}), P_0)} \tag{12}$$

comprising the latency τ_{DSP} [11] in DSP processing and the acoustical delay depending on the distance $|\mathbf{r}\text{-}\mathbf{r}_e|$ and the local speed of sound c_0, which is a function of the temperature field $T_A(\mathbf{r})$ and the static sound pressure P_0.

The (real) sound power $\Pi_L(f)$ radiated by the loudspeaker into the far-field can be calculated by multiplying the wave coefficients $\mathbf{C}_L(f)$ with its Hermitian transpose:

$$\Pi_L(f) = \frac{\mathbf{C}_L(f)\mathbf{C}_L^H(f)}{2\rho_0 c k^2}|U(f)|^2 \tag{13}$$

This sound power $\Pi_L(f)$ is a valuable metric for describing the global acoustic output of the loudspeaker by a single value. Still, it is also a convenient basis to estimate the mean sound pressure of the diffuse sound generated in a non-anechoic room if the reverberation time is known [11].

5. Time-variant distortion

The gray box model from **Figure 2** describes the time-variant distortion spectrum $P_v(f,\mathbf{r}|t)$ at any point $\mathbf{r}$ in the sound-field as

$$P_V(f, \mathbf{r}|t) = (H_V(f|t) - 1)H_L(f, \mathbf{r})U(f) \tag{14}$$

Using the spatial transfer $H_L(f,\mathbf{r})$, and the input spectrum $U(f)$, and the time-variant transfer function $H(f|t)$, which can be identified as the ratio

$$H_V(f|t) \approx \frac{H(f, \mathbf{r}|t)}{H(f, \mathbf{r}|t_0)} \tag{15}$$

of two spatial transfer functions $H(f,\mathbf{r}|t_0)$ and $H(f,\mathbf{r}|t)$ measured on the same loudspeaker unit under identical measurement conditions (environment, evaluation point $\mathbf{r}$) at a reference time t_0 and a later evaluation time t. The reference measurement at t_0 assesses the loudspeaker under climatized standard conditions using a small stimulus generating negligible heating and nonlinear distortion. The subsequent measurement at time t can be performed with any stimulus providing sufficient excitation of the loudspeaker. This measurement requires no scanning process, and the calculated time-variant transfer function $H_v(f|t)$ is independent of the choice of the evaluation point $\mathbf{r}$. Placing the microphone in the near-field ensures a good SNR.

This model is able to predict the amplitude compression at any point $\mathbf{r}$ in the sound-field defined in agreement with IEC standard 60268–21 [11] in decibel as

$$C_{AC}(f, \mathbf{r}|t) = -20\lg(|H_V(f|t)|)dB \tag{16}$$

and the phase deviation:

$$\Delta\varphi(f, \mathbf{r}|t) = \arg(H_V(f|t)) \tag{17}$$

The voice coil heating in professional stage loudspeakers can cause significant amplitude compression (up to 6 dB) in the output signal. Fatigue and climate changes can also shift the resonance frequencies of modal cone vibrations, causing more than 90-degree phase deviation. Those variations can impair the intended superposition of multiple loudspeakers' output in spatial sound applications.

6. Nonlinear distortions

The regular nonlinear distortions found in the sound pressure output $p_N(t,\mathbf{r})$ are symptoms of loudspeaker nonlinearities modeled by subsystems N_I and $N_D(\mathbf{r})$ shown in **Figure 2**. The input signal u strongly influences the generation process and the spectral and temporal properties of the nonlinear distortion [22].

A typical audio signal (e.g., music) has a dense excitation spectrum, as shown in **Figure 11**, which makes separating the nonlinear distortion p_N in the sound pressure output p more difficult. An adaptive linear filter can model the linear and time-variant components $p_L + p_V$ in the output [27]. The difference signal $e(t)$ between the measured and the modeled signal comprises nonlinear distortion and noise.

As shown in **Figure 11**, a sparse multi-tone complex is a stimulus able to represent typical program material such as music and speech by having similar properties such as spectral distribution and crest factor. This stimulus has pseudo-random properties generated by a standardized algorithm [11] to ensure reproducible and comparable test results. The excitation tones are not dense but sufficiently activate harmonics, intermodulation, and other nonlinear distortion components, which can easily be detected and separated from the fundamental response in the spectrum.

The prevalent measurement technique uses a single tone stimulus with a constant or varying excitation frequency f_e (e.g., sinusoidal chirp [11]). The harmonic components generated at multiple frequencies nf_e with $n = 2, 3, 4$ can be easily separated from the fundamental part at f_e. This measurement technique has a long tradition and is simple but has a significant drawback: It does not consider the intermodulation distortion generated by multiple tones and music.

The measurement technique presented in the following section can also be applied to a burst signal, two-tone signal, white or pink noise, and other input signals.

6.1 Nonlinear distortion in 3D space

A comprehensive measurement of the nonlinear distortion in the 3D space requires near-field scanning providing the distortion spectrum $P_N(f,\mathbf{r}_k)$ at the grid points $\mathbf{r}_k \in S_r$. The small distance between the microphone and loudspeaker ensures sufficient SNR to cope with noise. The measurement performed at high amplitudes can be integrated into the scanning process for spatial transfer function $H_L(f,\mathbf{r})$ measured at low amplitude (see Section 4).

Applying the spherical wave expansion to the measured distortion spectrum $P_N(f,\mathbf{r}_k)$ gives the optimal coefficients

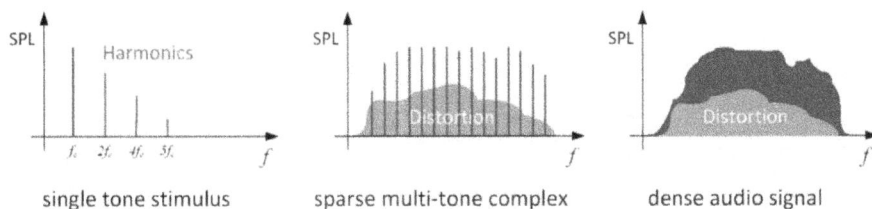

Figure 11.
Spectra of reproduced test stimuli used for nonlinear distortion measurement.

$$\mathbf{C}_N(f) = \underset{\mathbf{C}}{\mathrm{argMIN}} \sum_{k=1}^{K_r} |P_N(f, \mathbf{r}_k) - \mathbf{C}(f)\mathbf{B}_{out}(f, \mathbf{r}_k)|^2 \qquad (18)$$

The coefficients in vector $\mathbf{C}_N(f)$ allow extrapolation of the distortion to any point $\mathbf{r}$ outside the scanning surface:

$$P_N(f, \mathbf{r}) = \mathbf{C}_N(f)\mathbf{B}_{out}(f, \mathbf{r}) \qquad (19)$$

However, there is a significant difference between the nonlinear coefficients $\mathbf{C}_N(f)$ and the linear coefficients $\mathbf{C}_L(f)$ discussed in Section 4. The linear coefficients $\mathbf{C}_L(f)$ are parameters of a linear system. They can be identified with any broad-band stimulus and used to transfer another input signal into the sound-field, including music and speech. The nonlinear coefficients $\mathbf{C}_N(f)$ describes the results (distortion) of loudspeaker nonlinearities that depend on the particular stimulus [22]. The sound power spectrum calculated as

$$\Pi_N(f) = \frac{\mathbf{C}_N(f)\mathbf{C}_N^H(f)}{2\rho_0 ck^2}|U(f)H_V(f, |t)|^2 \qquad (20)$$

is a valuable global metric to assess the nonlinear distortion radiated by the loudspeaker in all directions.

6.2 Equivalent input distortion

The standard IEC 60268–21 calculates the equivalent input distortion (EID) for a single point measurement $\mathbf{r}_k$ by a simple approximation [28]

$$U_I(f, \mathbf{r}_k) = \frac{P_N(f, \mathbf{r}_k)}{H_V(f, |t)H_L(f, \mathbf{r}_k)} \qquad (21)$$

using the time-variant transfer functions $H_V(f|t)$ and spatial transfer function $H_L(f|\mathbf{r})$. This inverse filtering transforms the sound pressure distortion $p_N(\mathbf{r}_k)$ into virtual input signal u' $(\mathbf{r}_k)$, as illustrated in **Figure 12**.

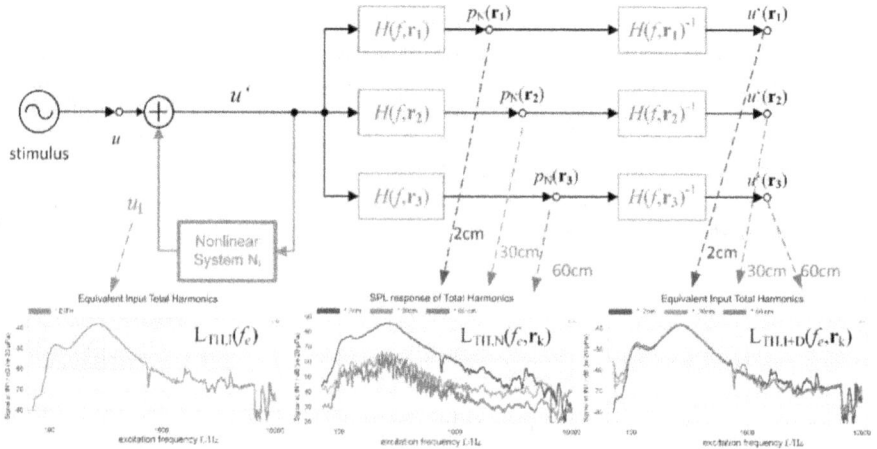

Figure 12.
Block diagram illustrates the calculation of equivalent input distortion (EID) by applying inverse filtering (right) or optimal estimation (left) based on three sound pressure measurements in the near-field (middle).

The lower middle panel in **Figure 12** shows the total harmonic distortion as an absolute SPL frequency response $L_{TH,N}(f_e,\mathbf{r})$ measured at three different distances $\mathbf{r}_k$ in an office room (in-situ). The near-field measurement at 2 cm provides a relatively smooth curve, while the 30 and 60 cm measurements have a lower SPL and are affected by room reflections. The filtering of the sound pressure signals p $(\mathbf{r}_k)$ with the inverse transfer function $H(f,\mathbf{r}_k)^{-1}$ generates a voltage signal u' $(\mathbf{r}_k)$ with the total harmonics level $L_{TH,\,I\,+\,D}(f_e,\mathbf{r}_k)$ on the lower right-hand side in **Figure 12**. This filtering removes the peaky curve shape caused by the room reflections, and the three curves become virtually identical between 100 Hz and 1 kHz. However, noise corrupts the measurement at low frequencies, and the distributed distortion p_D causes minor deviations above 800 Hz.

Those artifacts in the equivalent input distortion (EID) can be removed by minimizing the mean squared error between the estimated and the measured nonlinear distortion spectrum at the scanning points $\mathbf{r}_k$ with $k = 1,..,$ K_r and $K_r \geq 1$:

$$U_I(f) = \underset{U_{EID}}{\mathrm{argMIN}} \sum_{k=1}^{K_r} |H_V(f,|t)H_L(f,\mathbf{r}_k)U_I(f) - P_N(f,\mathbf{r}_k)|^2 \qquad (22)$$

This fitting provides the voltage level response $L_{TH,I}(f)$ on the left-hand side in **Figure 12**, representing the EID.

Figure 13 shows the equivalent input distortion spectrum $U_I(f)$ generated by multi-tone stimuli with a different spectral shaping to represent typical test signals and selected audio material. All the stimuli have the same RMS value. Cello music provides the highest low-frequency components, generating the highest voice coil displacement and harmonic components at 500 Hz. Pink noise and IEC noise [11], representing typical program material, cause harmonic and intermodulation distortion at the same SPL over a wide frequency band. The nonlinear distortion rise to higher frequencies for voice and white noise stimuli.

The EID spectrum $U_I(f)$ at the input of the loudspeaker can also be easily transferred to at any point $\mathbf{r}$ in the 3D space by applying linear filtering:

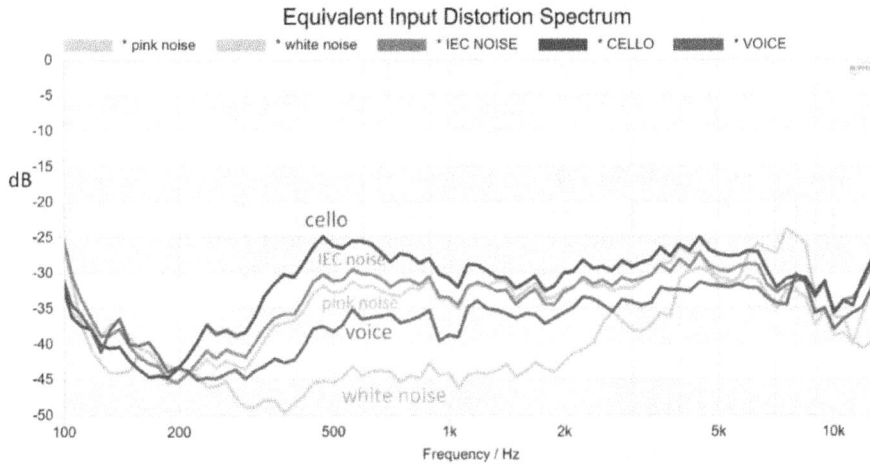

Figure 13.
Relative equivalent input distortion $L_I(f)$ measured with various broad-band stimuli at the same RMS input voltage.

$$P_I(f, \mathbf{r}) = H_V(f|t)H_L(f, \mathbf{r})U_I(f)$$
$$= \mathbf{C}_L(f)\mathbf{B}_{\text{out}}(f, \mathbf{r})H_V(f|t)U_I(f) \tag{23}$$

The sound power spectrum $\Pi_I(f)$ of the equivalent input distortion radiated into the far-field can be similarly calculated as the linear power $\Pi_L(f)$ in Eq. (13) by using the same wave coefficients $\mathbf{C}_L(f)$ of the linear wave modeling:

$$\Pi_I(f) = \frac{\mathbf{C}_L(f)\mathbf{C}_L^H(f)}{2\rho_0 ck^2}|U_I(f)H_V(f, |t)|^2 \tag{24}$$

The transfer functions $H_L(f,\mathbf{r})H_v(f|\mathbf{r})$ shape the spectral components of equivalent input distortion and the input stimulus in the same way. Thus, the ratio between distortion and linear signal part is identical in the voltage, sound pressure at any point $\mathbf{r}$, and power output:

$$\frac{|U_I(f)|}{|U(f)|} = \frac{|P_I(f, \mathbf{r})|}{|P_L(f, \mathbf{r})|} = \sqrt{\frac{|\Pi_I(f)|}{|\Pi_L(f)|}} \tag{25}$$

This fact simplifies the distortion measurement and motivates the definition of relative distortion metrics discussed in Section 6.4. Furthermore, nonlinear control techniques [17] that cancel the EID at the loudspeaker input by synthesized compensation signal can reduce the sound pressure distortion $P_I(f,\mathbf{r})$ everywhere in the 3D space.

6.3 Distributed nonlinear distortion

The distributed nonlinear distortion $p_D(\mathbf{r})$ introduced in Section 2 is the remaining distortion part in the sound-field that EID cannot represent:

$$P_D(f, \mathbf{r}_k) = (P_N(f, \mathbf{r}_k) - H_V(f|t)H_L(f, \mathbf{r}_k)U_I(f))$$
$$= \mathbf{C}_D(f)\mathbf{B}_{\text{out}}(f, \mathbf{r}_k)H_V(f, |t)U(f) \tag{26}$$

Eq. (26) uses the basic functions $\mathbf{B}_{\text{OUT}}(f, \mathbf{r})$ from Eq. (6) for the spherical wave expansion but determines the coefficients $\mathbf{C}_D(f)$ as:

$$\mathbf{C}_D(f) = \arg\underset{\mathbf{C}}{\text{MIN}} \sum_{k=1}^{K_r} |P_D(f, \mathbf{r}_k) - \mathbf{C}(f)\mathbf{B}_{\text{out}}(f, \mathbf{r}_k)|^2 \tag{27}$$

The residual error in Eq. (27) can be used to find the maximum order N of the wave expansion, as discussed in Section 4. The symmetry properties of the particular loudspeaker are also valuable for minimizing the scanning effort.

The coefficients $\mathbf{C}_D(f)$ provide the sound power spectrum $\Pi_D(f)$ of the distributed nonlinear distortion radiated into the far-field as:

$$\Pi_D(f) = \frac{\mathbf{C}_D(f)\mathbf{C}_D^H(f)}{2\rho_0 ck^2}|U(f)H_V(f, |t)|^2 \tag{28}$$

The distributed distortion can be ignored if the sound power $\Pi_D(f)$ is smaller than one-tenth of the EID sound power $\Pi_I(f)$. Then a single test in the near-field of the loudspeaker is sufficient to measure the dominant EID and predict the total distortion p_N in the 3D space.

6.4 Relative distortion metrics

This section introduces metrics that simplify the interpretation of the distortion components. These equations use a symbol # as a placeholder for N, I, or D representing the total, equivalent input, or distributed distortion.

Comparing the spectral components at frequency f in the nonlinear distortion $P_\#(f,\mathbf{r})$ with the linear output signal $P_L(f,\mathbf{r})$ from Eq. (6) at the same point $\mathbf{r}$ leads to a spectral nonlinear distortion ratio (SNDR) defined in decibel as:

$$L_\#(f,\mathbf{r}) = 20\lg\left(\frac{|P_\#(f,\mathbf{r})|}{|P_L(f,\mathbf{r})|}\right)dB \quad \# \in \{N, I, D\} \tag{29}$$

The SNDR is usually negative and describes the SPL difference between the distortion and the linear component at the same spectral frequency f.

It is a proper physical metric for broad-band stimuli such as typical audio signals, noise, and other artificial test stimuli. It also applies to sparse multi-tone stimuli with a resolution smaller than one-third octave by using $P_\#(f_i,\mathbf{r})$ in the nominator of Eq. (29) and the fundamental component $P_L(f_j, \mathbf{r})$ in the denominator with the smallest frequency difference $|f_i\text{-}f_j|$ for each spectral distortion component.

However, SNDR) is less useful for sinusoidal stimuli generating only a single tone with constant or varying excitation frequency (e.g., chirp) because the harmonics have a significant spectral distance to the fundamental.

An alternative approach considers the total energy ratio between the nonlinear distortion $P_\#$ and the linear output signal P_L for a particular stimulus. It leads to the total distortion ratio (TDR) defined in percent as:

$$R_\#(\mathbf{r}) = \sqrt{\frac{\int |P_\#(f,\mathbf{r})|^2 df}{\int |P_L(f,\mathbf{r})|^2 df}}100\% \quad \# \in \{N, I, D\} \tag{30}$$

This metric can be applied to all kinds of stimuli but is very popular for the total harmonic distortion THD measured with a single tone and plotted versus the excitation frequency f_e. This metric does reveal the spectral distribution of the nonlinear distortion (second, third, and higher-order harmonics).

Referring the nonlinear sound power spectrum $\Pi_\#(f)$ to the linear sound power $\Pi_L(f)$ in Eq. (13) provides a sound power distortion ratio (SPDR):

$$R_{\Pi,\#} = \sqrt{\frac{\int \Pi_\#(f)df}{\int \Pi_L(f)df}}100\% \quad \# \in \{N, I, D\} \tag{31}$$

For a multi-tone stimulus representing typical program material (IEC 60268–21), the SPDR becomes an essential, single-value characteristic for the assessment of the audio quality in a global sense.

The spectral equivalent input distortion ratio (SEIDR) defined in decibel as

$$L_I(f) = 20\lg\left(\frac{|U_I(f)|}{|U(f)|}\right)dB = L_I(f,\mathbf{r})\approx L_N(f,\mathbf{r}) \tag{32}$$

compares the spectral components of distortion $U_I(f)$ with the input signal $U(f)$. The metric $L_I(f)$ is identical with the metric $L_I(f, \mathbf{r})$, assessing the EID at any point $\mathbf{r}$ in the sound-field. It is a valid approximation for the total distortion metric $L_N(f, \mathbf{r})$ if the distributed distortion $P_D(f,\mathbf{r})$ is negligible.

7. Abnormal distortion

Loudspeaker defects such as voice coil rubbing, mechanical vibrations of loose parts, air turbulences, and other irregular nonlinear dynamics that are neither intended nor considered in the design can generate particular distortion that can significantly degrade the audio quality. A loudspeaker generating abnormal distortion, usually called "rub & buzz" should not be shipped to a customer!

Modern measurement techniques exploit unique features of abnormal distortion. Time-analysis applied to a distorted single-tone stimulus reveals a complex fine structure comprising spikes, transients, and noise-like patterns [29]. Contrary to the harmonic and intermodulation distortion discussed in Section 6, the abnormal distortions cover the entire audio band. However, they have a low RMS value, are usually close to the noise floor, and thus require a near-field measurement. Spherical wave expansion or averaging over multiple periods removes the random features of the abnormal distortion.

The IEC standard 60268–21 [11] recommends a chirp stimulus at varying excitation frequency f_e and a high-pass tracking filter with a cut-off frequency $f_c > n_{co}f_e$ to separate the abnormal distortion in the measured sound pressure signal $p(t)$. The factor n_{co} for the cut-off frequency f_c (typical value $n_{co} = 10$) depends on the excitation frequency f_e, the transducer type, and properties of potential defects. The optimal value for n_{co} can be determined by maximizing the crest factor $C_{ID}(\mathbf{r})$ defined according to IEC 60268–21 [11] as the ratio between peak and RMS values of the high-pass filtered signal p_{ID} as:

$$C_{ID}(\mathbf{r}) = 10\lg\left(\frac{MAX_t^{t+T}|p_{ID}(t,\mathbf{r})|^2}{\frac{1}{T}\int_t^{t+T}p_{ID}(t,\mathbf{r})^2 dt}\right)dB \tag{33}$$

The crest factor $C_{ID}(\mathbf{r})$ is independent of the spectral energy but describes the impulsiveness of the abnormal distortion considering the phase relationship between the spectral components. A high crest factor is a unique symptom of abnormal distortion, while the crest factor of the fundamental, regular nonlinear distortions or electronic noise is typically below 12 dB.

This fact initiated the measurement of the impulsive distortion (ID) defined in IEC 60268–21 as a peak level in decibel as

$$L_{ID}(f_e,\mathbf{r}_k) = 20\lg\left(\frac{MAX_t^{t+T}|p_{ID}(t,\mathbf{r}_k)|}{p_{ref}}\right)dB \tag{34}$$

Using a peak found over a period length T in the nominator in Eq. (33) and normalized by reference sound pressure p_{ref}. This peak level $L_{ID}(f_e,\mathbf{r}_k)$ is a helpful metric for finding the most critical excitation frequency f_{ID} and a scanning point $\mathbf{r}_{ID} \in S_r$ at the nearest position to the source (e.g., rattling), generating impulsive distortion with $C_{ID}(f) > 12$ dB. The maximum value found under the condition

$$L_{IDmax} = L_{ID}(f_{ID},\mathbf{r}_{ID}) = \underset{\forall f_e}{MAX}\left(\underset{\forall \mathbf{r}_k \in S_r}{MAX}(L_{ID}(f_e,\mathbf{r}_k))|C_{ID}(f) > 12dB\right) \tag{35}$$

is the basis for calculating the maximum impulsive distortion ratio (IDR) defined according to IEC 60268–21 [11] as

$$L_{IDR} = L_{ID}(f_{ID},\mathbf{r}_{ID}) - L_{REF} \tag{36}$$

using a reference sound pressure level L_{REF} measured at the standard evaluation point (on axis, $r = 1$ m) or a scanning point $\mathbf{r}_k$ generating the largest SPL value:

$$L_{REF} = 10\lg\left(\frac{1}{Tp_{ref}^2}\underset{\forall \mathbf{r}_k \in S_r}{MAX}\int\limits_{t}^{t+T} p(t,\mathbf{r}_k)^2 dt\right)dB \qquad (37)$$

Those metrics compared with meaningful limits for passing or failure are essential for the quality control of loudspeakers in manufacturing and maintenance.

8. External noise

The SNR in decibel is defined as

$$R_{SNR} = L_{REF} - L_N \qquad (38)$$

using reference SPL L_{REF} from Eq. (37) and a noise SPL L_N. The stationary noise caused by the microphone and other electronic parts can be measured with a muted stimulus in a single test at any point $\mathbf{r}$. The instantaneous SNR can be used to validate the distortion ratios TDR in Eq. (30) and IDR in Eq. (36) to remove invalid data.

9. Metrics for sound zones

Audio quality assessment, loudspeaker diagnostics, and active sound-field control require metrics that assess the properties of the sound-field at a specific listening point described by a probability $f_L(\mathbf{r})$ of the ear position. The mean sound power found in such a listening zone is a less suitable metric because the listener evaluates the local sound pressure. It is more appropriate to assess the mean and the variance of the perceptual attributes (e.g., loudness) or related physical metrics (e.g., SPL) over the listening zone [30] considering the probability of the ear positioning as a weighting function $f_L(\mathbf{r})$. This approach is used in IEC 60268–21 [11] for defining a mean SPL over an acoustical zone, but it can easily be applied to the nonlinear distortion metrics in Eqs. (29) and (30). The variance and the maximum deviation from the mean value are also valuable characteristics of the sound zone.

10. Maximum SPL output

The maximum sound pressure output (max SPL) rated according to IEC standard 60268–21 [11] plays a primary role in adjusting the amplitude of the test stimulus in output-based testing. The max SPL can be used to calibrate any input channel (digital, analog) in passive and active systems and provides a maximum input RMS value u_{max}, depending on the selected input channel, gain control, amplification, and applied signal processing. The amplitude compression $C_{AC}(f)$ from Eq. (16), the sound power distortion ratio R_{TIN} from Eq. (31), and the maximum impulsive distortion ratio R_{IDR} from Eq. (36) are essential criteria for rating max SPL considering the particularities of the target applications.

11. Conclusions

Acoustical measurement in the near-field of the loudspeaker can provide much of the relevant information required for designing and assessing spatial sound control applications. The spatial transfer function $H_L(f,\mathbf{r})$ expressed as a spherical wave expansion provides accurate sound pressure amplitude and phase information at any point $\mathbf{r}$ in the near and far-field. The spatial scanning effort depends on the particular loudspeaker and can be significantly minimized by considering the symmetry of the loudspeaker. In practice, the spatial transfer function $H_L(f,\mathbf{r})$ scanned on a prototype can be applied to other units of the same type as long as the loudspeaker geometry does not change much.

The time-variant transfer function $H_v(f|t)$ represents changes in the material caused by heating, aging, fatigue, and production variability. No scanning is required to measure the transfer function $H_v(f|t)$ and the equivalent input distortion $U_I(f)$, ignoring the distributed nonlinear distortion p_D. Such an approximation is valid for most loudspeakers used in spatial sound applications and can be verified by scanning the nonlinear distortion in the near-field of the loudspeaker. All time-variant and nonlinear signal distortion can be extrapolated to any point in the 3D space using spherical wave expansions.

The multi-tone complex is a valuable artificial stimulus that can simplify the interpretation of the amplitude compression and the nonlinear distortion. The sinusoidal chirp is required to measure the impulsive distortion ratio, a sensitive characteristic for detecting loudspeaker defects and abnormal behavior degrading the audio quality.

An anechoic room is usually not required for performing the essential loudspeaker measurements at superior accuracy.

The methods for measuring loudspeaker characteristics presented in this chapter are compliant with modern international loudspeaker standards. They are the basis for simplifying the numerical simulation of sound-field control and selecting optimal hardware components offering a maximum performance-cost ratio.

Author details

Wolfgang Klippel
KLIPPEL GmbH, Dresden, Germany

*Address all correspondence to: wklippel@klippel.de

IntechOpen

References

[1] Van Veen BD, Buckley KM. Beamforming: A versatile approach to spatial filtering. IEEE ASSP Magazine. 1988;5(2):4-24

[2] Berkhout AJ, Vries DD, Vogel P. Acoustical control by wave field synthesis. Journal of the Acoustical Society of America. 1993;93:2764-2778

[3] Gerzon MA. Ambisonics in multi-channel broadcasting and video. Journal of the Audio Engineering Society. 1985; 33(11):859-871

[4] Poletti M. Three-dimensional surround sound systems based on spherical harmonics. Journal of the Audio Engineering Society. 2005;53(11): 1004-1025

[5] Betlehem T, Zhang W, Poletti M, Abhayapala T. Personal sound zones: Delivering Interface-free audio to multiple listeners. IEEE Signal Processing Magazine. 2015;32:81-91

[6] Zotter F. Analysis and Synthesis of Sound Radiation with Spherical Arrays [Dissertation]. Austria: University of Music and Performing Arts; 2009

[7] Vries DD. Sound reinforcement by wave field synthesis: Adaptation of the synthesis operator to the loudspeaker directivity characteristics. Journal of the Audio Engineering Society. 1996; 44(12):1120-1131

[8] Ahrens J, Spors S. An analytical approach to 2.5 D sound field reproduction employing linear distributions of non-omnidirectional loudspeakers. In: Proc. IEEE Int. Conf. Acoust. Speech and Signal Process. (ICASSP). 2010. pp. 105-108. DOI: 10.1109/ICASSP15600.2010

[9] Koyama S, Furuya K, Hiwasaki Y, Haneda Y. Sound field reproduction method in Spatio-temporal frequency domain considering directivity of loudspeakers. In: 132nd Convention of the Audio Eng. Soc., Budapest, Paper 8664. 2012. Available from: http://www.aes.org/e-lib/browse.cfm?elib=16302

[10] Poletti MAA, Betlehem T, Abhayapala THD. Higher-order loudspeakers and active compensation for improved 2D sound field reproduction in rooms. Journal of the Audio Engineering Society. 2015;63(1/2):31-45. DOI: 10.17743/jaes.2015.0003

[11] Sound System Equipment – Part 21. Acoustical (Output-Based) Measurements. Standard of International Electrotechnical Commission IEC 60268–21; 2018

[12] Choi J, Kim Y, Ko S. Near and far-field control of focused sound radiation using a loudspeaker Array. In: 129th Convention of the Audio Eng. Soc., San Francisco, Paper 8198. 2010. Available from: http://www.aes.org/e-lib/browse.cfm?elib=15620

[13] Ma X et al. Nonlinear distortion reduction in sound zones by constraining individual loudspeaker control effort. Journal of the Audio Engineering Society. 2019;57(9):641-654

[14] Cobianchi M, Mizzoni F, Uncini A. Polar measurements of harmonic and multitone distortion of direct radiating and horn loaded transducers. In: 134th Convention of the Audio Eng. Soc., Rome, Paper 8915. 2013. Available from: http://www.aes.org/e-lib/browse.cfm?elib=16815

[15] Olsen M, Møller MB. Sound zones: On the effect of ambient temperature variations in feed-forward systems. In: 142nd Convention of the Audio Eng. Soc., Berlin, Paper 9806. 2017. Available from: http://www.aes.org/e-lib/browse.cfm?elib=18680

[16] Pedersen KM. Thermal overload protection of high-frequency loudspeakers [Rep. of final year dissertation]. UK: Salford University; 2002

[17] Klippel W. Loudspeaker and headphone design approaches enabled by adaptive nonlinear control. Journal of the Audio Engineering Society. 2020; 68(6):454-464. DOI: 10.17743/jaes.2020.0037

[18] Klippel W. Nonlinear Modeling of the heat transfer in loudspeakers. Journal of the Audio Engineering Society. 2004;52(1/2):3-25

[19] Klippel W. Mechanical fatigue and load-induced aging of loudspeaker suspension. In: 131st Convention of Audio Eng. Soc., New York, paper 8474. 2011. Available from: http://www.aes.org/e-lib/browse.cfm?elib=16000

[20] Klippel W, Schlechter J. Distributed mechanical parameters of loudspeakers, part 1: Measurements. Journal of the Audio Engineering Society. 2009;57(7/8):500-511

[21] Sound system equipment - Part 22. Electrical and Mechanical Measurements on Transducers. Standard of International Electrotechnical Commission, IEC 60268–21; 2020

[22] Klippel W. Loudspeaker nonlinearities – Causes parameters, symptoms. Journal of the Audio Engineering Society. Oct 2006;54(10): 907

[23] Sound System Equipment – Electro-acoustical Transducers – Measurement of Large Signal Parameters, Standard of International Electrotechnical Commission, IEC 62458; 2010

[24] Klippel W, Bellmann C. Holographic nearfield measurement of loudspeaker directivity. In: 141st

Convention of the Audio Eng. Soc., Los Angeles, Paper 9598. 2016. Available from: http://www.aes.org/e-lib/browse.cfm?elib=18402

[25] Williams EG. Fourier Acoustics – Sound Radiation and Nearfield Acoustical Holography. London: Academic Press; 1999

[26] Melon M et al. Comparison of four subwoofer measurement techniques. Journal of the Audio Engineering Society. 2007;55(12):1077-1091

[27] Klippel W, Irrgang S. Audio system evaluation with music signals. In: AES International Conference on Automotive Audio, San Francico, Paper P4–2. 2017. Available from: http://www.aes.org/e-lib/browse.cfm?elib=19196

[28] Klippel W. Measurement and application of equivalent input distortion. Journal of the Audio Engineering Society. 2004;52(9): 931-947

[29] Klippel W, Seidel U. Measurement of impulsive distortion, rub and buzz and other disturbances. In: 114th Convention of Audio Eng. Soc, Paper 5734. 2003. Available from: http://www.aes.org/e-lib/browse.cfm?elib=12550

[30] IEC 62777. Quality Evaluation Method for the Sound Field of Directional Loudspeaker Array System. Standard of the International Electrotechnical Commission. 2016

Perspective Chapter: Modern Acquisition of Personalised Head-Related Transfer Functions – An Overview

Katharina Pollack, Wolfgang Kreuzer and Piotr Majdak

Abstract

Head-related transfer functions (HRTFs) describe the spatial filtering of acoustic signals by a listener's anatomy. With the increase of computational power, HRTFs are nowadays more and more used for the spatialised headphone playback of 3D sounds, thus enabling personalised binaural audio playback. HRTFs are traditionally measured acoustically and various measurement systems have been set up worldwide. Despite the trend to develop more user-friendly systems and as an alternative to the most expensive and rather elaborate measurements, HRTFs can also be numerically calculated, provided an accurate representation of the 3D geometry of head and ears exists. While under optimal conditions, it is possible to generate said 3D geometries even from 2D photos of a listener, the geometry acquisition is still a subject of research. In this chapter, we review the requirements and state-of-the-art methods for obtaining personalised HRTFs, focusing on the recent advances in numerical HRTF calculation.

Keywords: head-related transfer functions, spatial hearing, acoustic measurement, numerical calculation, localisation

1. Introduction

Head-related transfer functions (HRTFs) describe the filtering of the acoustic field produced by a sound source arriving at the listener's ear. The filtering is the effect of the interaction of the sound field with the listener's anatomy and has various properties. First, the incoming sound wave arrives at the ipsilateral pinna, i.e., the ear closer to the sound source, and then at the contralateral ear, i.e., the ear away from the sound source. This time difference between ipsilateral and contralateral ear is usually described as the interaural time difference (ITD). Second, larger anatomical structures, i.e., torso, shoulders and head, affect frequencies up to 3 kHz in a comparatively trivial way. As the listener's torso and head shadow the sound wave arriving at the contralateral ear, interaural level differences (ILDs) arise. Third, the incoming sound is filtered in a complex way by the shape of the listener's pinnae. These monaural time-frequency-filtering effects become especially important for higher frequency regions (above approximately 4 kHz) and sound directions inducing the same ITDs and ILDs [1–6]. Humans have learned to interpret this acoustic filtering to span an auditory space as an internal model of their natural environment [7]. Because the pinna shape is unique for every person, HRTFs are considered listener-specific [8–10], similar to a fingerprint [1–6]. With an individually fitted HRTF dataset, it is possible for a person to perceive sounds (in a virtual environment) via headphones as if the sounds would originate from their (physical) position around the listener.

Both interaural and monaural features for a single sound direction can be represented by a binaural HRTF pair [11]. In signal processing terms, a binaural HRTF pair can be described as

$$HRTF_L(\mathbf{x}^*, f, s) = \frac{p_L(\mathbf{x}^*, f, s)}{p_0(0, f)}$$

$$HRTF_R(\mathbf{x}^*, f, s) = \frac{p_R(\mathbf{x}^*, f, s)}{p_0(0, f)}$$

(1)

where p_L and p_R describe the sound pressure at a position inside the left and right ear, respectively (typically the entrance of the left and right ear canal or a position close to the eardrum), $\mathbf{x}^*$ describes the sound-source position (i.e., distance and direction), f describes the frequency and s the listener's geometry, emphasising the listener-specificity of HRTFs. p_0 describes the reference sound pressure, which is usually the pressure measured at the position of the midpoint of right and left ear *without* the head being present.

There are several options to set a specific coordinate system to systematically describe directions for HRTFs. From the physical perspective, the *spherical* coordinate system is a natural choice; in that case, the origin of the system is placed inside the listener's head at the midpoint between left and right ear and the direction is described by azimuth and elevation angles, see **Figure 1a**. In this system, one can intuitively define the two main planes: The eye-level horizontal plane, i.e., all directions with the elevation angle of zero, and the median plane, i.e., all directions with the azimuth angle of zero. The eye-level horizontal plane is also called Frankfurt plane and can be anatomically defined as the plane connecting the lowest part of the listener's orbital cavity and the highest part of the bony ear canal (meatus acusticus externus osseus). This spherical coordinate system resembles a *geodesic* representation widely used in physics, with the poles located at the top and bottom. An alternative system that is more relevant from the auditory perspective is given by the *interaural-*

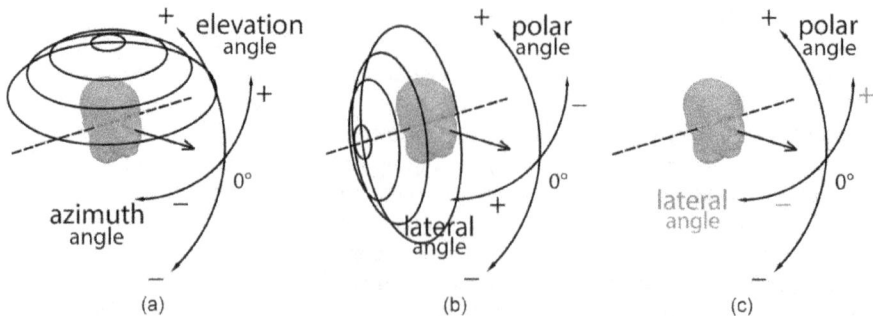

Figure 1.
Coordinate systems typically used in the HRTF acquisition and representation. The dashed line represents the interaural axis, and the arrow represents the viewing direction. (a) Spherical coordinate system with the azimuth and elevation angles. (b) Simple interaural-polar coordinate system with the lateral and polar angles obtained by rotation the poles of the spherical system. (c) Modified interaural-polar coordinate system with the lateral and polar angles corresponding to the azimuth angle in the horizontal plane and the elevation angle in the median plane.

polar coordinate system. This system is shown in **Figure 1b** and can be constructed by rotating the poles of the spherical system to the interaural axis, i.e., the axis connecting the two ears. A sound direction is then described by the lateral angles (along the horizontal plane) and polar angles (along the median plane). The poles are then located on the left and right sides of the listener. This simple interaural-polar coordinate system was used in various psychoacoustic studies, e.g., [12, 13], and has the disadvantage that the lateral angle does not correspond to the azimuth angle. **Figure 1c** shows the *modified* version of the interaural-polar coordinate system, which does not have this disadvantage. Here, the sign of the lateral angle is flipped, i.e., in the coordinate system, the positive lateral angles are used for sounds located on the left side of the listener. This transformation to a left-handed coordinate system has the advantage of having the lateral angle corresponding to the azimuth angle for all sources placed in the horizontal plane, and the polar angle corresponding to the elevation angle for all sources placed in the median plane. Thus, the modified interaural-polar coordinate system offers a better link between the psychoacoustic research and audio engineering. In that system, the lateral angle ranges from −90° (right ear) over 0° (front) to 90° (left ear), and the polar angle ranges from −90° (bottom) over 0° (front) and 90° (up) to 180° (back) and 270° (bottom again).

The understanding of these coordinate systems is important because state-of-the-art acquisitions and representations of HRTFs utilise those systems. For example, **Figure 2** shows HRTFs along the Frankfurt and the median plane. These various coordinate systems are used in HRTF visualisation, in various HRTF-related software packages such as the SOFA toolbox [15], and in auditory modelling, e.g., the Auditory Modelling Toolbox (AMT) [16, 17].

HRTF acquisition can be classified into three categories: acoustic measurement, numerical calculation, and personalisation [18].

The acoustic measurement is traditionally designed as the measurement of the impulse response between source and receiver in an anechoic or semianechoic chamber, describing the transmission path from a sound source to the ear [11, 19]. A comprehensive review of the established state-of-the-art acoustic techniques to measure HRTFs can be found in [20]. Thus, in this chapter, Section 3, we only briefly provide an overview of the traditional acoustic HRTF measurement approaches, highlight some of their differences and new trends and focus on the requirements for the acoustic measurement.

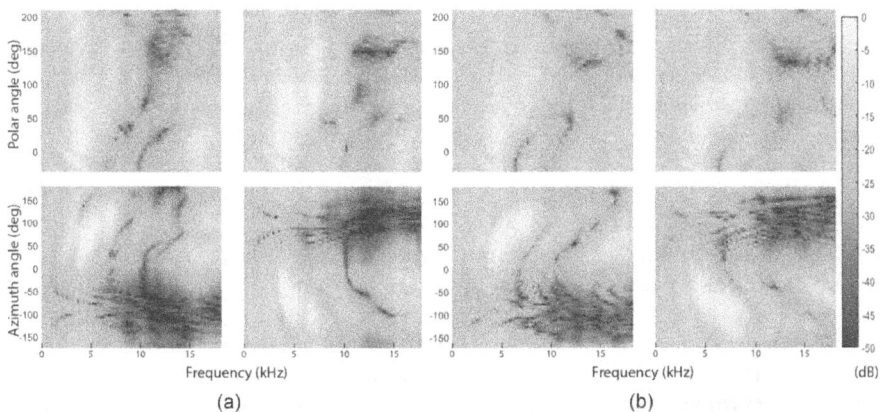

Figure 2.
HRTF magnitude spectra for the listeners (a) NH236 and (b) NH257, both from the ARI database [14]. Top: Spectra along the median plane. Bottom: Spectra along the eye-level horizontal plane. 0 dB corresponds to the maximum magnitude in each panel.

Numerical HRTF calculation simulates the acoustic measurement by considering a 3D representation of the listener's geometry and the positions of multiple external sound sources, for which the generated sound pressure at the entrance of the ear canal is calculated. This technique has become more popular and is the main focus of this chapter. To this end, in Section 4, we provide an overview of the principles of various numerical calculation approaches including a comparison of the mentioned methods.

Personalisation of HRTFs describes the process of adapting an existing set of generic data guided by listener-specific information, either with the help of objective or subjective personalisation method. The objective personalisation has been approached from two different domains: the geometric domain, in which listener-specific anthropometric data are measured and used to personalise a generic geometric model from which HRTFs are then simulated; or the spectral domain, in which a generic HRTF set is directly personalised based on listener-specific information. Examples for personalisation approaches include utilising frequency scaling [21], parametric modelling of peaks and notches [22], active shape modelling (ASM) [23], principal component analysis (PCA) in both geometric [24] and spectral domains [25–29], multiple regression analysis [30], independent component analysis (ICA) [31], large deformation diffeomorphic metric mapping (LDDMM) [25, 32], local neighbourhood mapping [33], neural networks [34–41] and linear combination of HRTFs [42]. Despite many efforts worldwide [43–46], the link between the morphology and HRTFs is not fully understood yet, mostly because of the high dimensionality of the problem. Most recent tools for studying that link are rooted in aligning high-resolution pinna representations to target representations facilitated with parametric pinna models [47, 48].

In the subjective personalisation, listeners are confronted with several sets of HRTFs and an algorithm (usually based on the evaluation of localisation errors, i.e., the difference between perceived and actual sound-source location) adapts the HRTF sets aiming at converging at listener-specific HRTFs [9, 49]. For an educated guess for the initial sets, anthropometric data can be used to pre-scale the HRTF sets, or the HRTF sets can be pre-selected via psychoacoustic models [50]. Clustering of the HRTF sets can further improve the relevance and reduce the duration of the personalisation procedure [49, 51].

All these methods aim at providing a specific quality in terms of acoustic and psychoacoustic properties. In the following section, we describe the acoustic properties and psychoacoustic requirements for human HRTFs, both of which lay the base for HRTF acquisition. Then, we briefly describe the most important requirements for the acoustic HRTF measurement, complementing the work of Li and Peissig [20]. Finally, we describe approaches for numeric HRTF calculation in greater detail.

2. Head-related transfer functions: acoustic properties and psychoacoustic requirements

In this section, we describe the acoustic properties of HRTFs and relate them to psychophysical properties of human hearing with the goal to derive the minimum requirements for sufficiently accurate HRTF acquisition by means of perception. We analyse spectral, temporal and spatial aspects of HRTFs and consider contributions of distinct parts of the human body to these aspects.

Humans can hear frequencies roughly between 20 Hz and 20 kHz, with frequencies at the lower end being perceived as vibrations or creaks, and with the upper end decreasing with age and duration of noise exposure [52]. From the psychoacoustic perspective, frequencies down to 90 Hz contribute to sound lateralisation, i.e., localisation on the interaural axis within the head [53], and up to 16 kHz to sound localisation, i.e., localisation outside the head [54], defining the smallest frequency range for the HRTF acquisition. **Figure 2** shows the amplitude spectra of a binaural HRTF pair of two listeners. For each listener, the left and right columns show HRTFs of the left and right ear, respectively. The top row shows the HRTFs along the median, i.e., for the lateral angle of zero, from the front, via up, to the back. The bottom row shows the HRTFs along the Frankfurt plane, i.e., the horizontal plane located at the eye level. **Figure 2** demonstrates that HRTFs vary across ears, frequency, sound-source positions and listeners. The bottom panels emphasise the difference between ipsilateral and contralateral ear, showing the dynamic range, especially for frequencies higher than 6 kHz.

Assuming the propagation medium is air and a sonic speed of 340 m/s, the human hearing frequency range translates to wavelengths approximately between 1.7 cm and 17 m, resulting in different body parts affecting HRTFs in different frequency regions. The reflections of the torso create spatial-frequency modulations in the range of up to 3 kHz [1]. This effect can be observed in the top row of **Figure 2**, in the form of elevation-dependent spectral modulations along the median plane [55, 56]. Another contribution comes from the head, which shadows frequencies above 1 kHz. This effect can be observed in both rows of **Figure 2**, with large changes in the spectra beginning at around 1 kHz [57]. A large contribution is that of the pinna: The resonances and reflections within the pinna geometry create spectral peaks and notches, respectively, in frequencies above 4 kHz [54]. This effect can be observed in the bottom row of **Figure 2**.

From the perceptual perspective, the quality of these HRTF spectral profiles is important in many processes involved in spatial hearing. For example, sound-localisation performance deteriorates when these spectral profiles are disturbed by means of introducing spectral ripples [58], reducing the number of frequency channels [59] or spectral smoothing [60]. From the acoustic perspective, these spectral profiles show modulation depths of up to 50 dB [11], defining the required dynamic range in the process of HRTF acquisition.

The temporal aspects of HRTF acquisition are shown in **Figure 3** as the head-related impulse responses (HRIRs), i.e., HRTFs in the time domain, of the same listeners as in **Figure 2**. There are a few things to consider. First, the minimum length of the measurement is bounded by the length of the HRIRs. Their amplitude decays within the first 5 ms, setting the requirement for the room impulse response during the measurements [61]. After the 5 ms, the HRIRs decay below 50 dB, setting the requirement on the broadband signal-to-noise ratio (SNR) of the measurements. Further, because of the human sensitivity to interaural disparities, HRTF acquisition

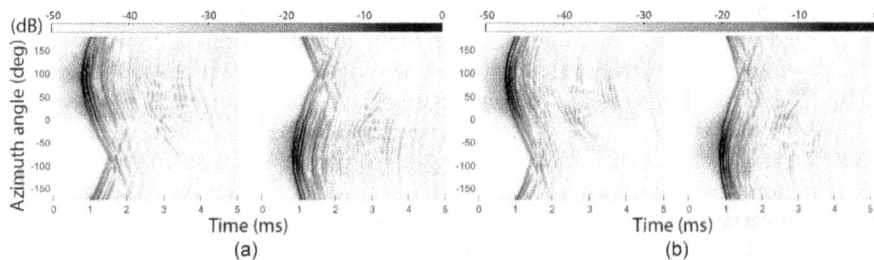

Figure 3.
*HRTF log-magnitudes in time domain along the eye-level horizontal plane for the same listeners as in **Figure 2**. Note the decay within the first 5 ms.*

also requires an interaural temporal synchronisation. While sound sources placed in the median plane cause an ITD of zero (theoretically, reached only for identical path lengths to the two ears), just small deviations from the median plane cause potentially perceivable non-zero ITDs. Human listeners can detect ITDs being as small as 10 μs [53, 62], defining the interaural temporal precision required in the HRTF acquisition process. The ITD increases with the lateral angle of the sound source, reaching its extreme values for sources placed near the interaural axis [63, 64]. The largest ITD depends on the distance between the listener's two ears, mostly being defined by the listener's head width and depth [65], reaching ITDs of up to ±800 μs. That ITD range translates to the sound's time of arrival (TOA) at an ear varying in the range of 1.6 ms, which needs to be considered in HRTF measurement by providing sufficient temporal space in the resulting impulse response.

HRTFs are continuous functions in space, even though, they are traditionally acquired for a finite set of spatial positions. From the *acoustic* perspective, assuming an HRTF bandwidth of 20 kHz, at least 2209 spatial directions are required to capture all spectro-spatial HRTF variations [66]. While this quite large number of spatial directions increases even further when considering multiple sound distances, it is in discrepancy with a smaller number of directions usually used in HRTF acquisition [11, 67–69]. One reason is the much smaller *perceptual* spatial resolution. From that perspective, the spatial resolution is limited by the ability to evaluate ITDs and changes in HRTF spectral profiles, both of which converge in the so-called minimum audible angles (MAAs). The MAA indicates the smallest detectable angle between two sound sources [70]. It depends on signal type [71, 72] and is minimal for broadband sounds [54, 73–75]. The MAA further depends on the direction of the source movement. Along the horizontal plane, the MAA can be as small as 1° for frontal sounds [76], increasing up to 10° for lateral sounds [77–79]. This translates to a high spatial-resolution requirement for frontal directions that can be relaxed with increasing lateral angle. Along the vertical planes, the MAA can be as low as 4° for frontal and rear sounds [76], increasing up to 20° for other sound directions [80]. Note that further relaxation of the requirement for spatial resolution can be achieved by using interpolation algorithms in the sound reproduction. For example, when using amplitude panning between the vertical directions [81], a resolution better than 30° does not seem to provide further advantages for localisation of sounds in the median plane [82]. Finally, when it comes to dynamic listening situations (involving listener or source movements), the MAAs further increase [83]. In order to account for sufficient spatial resolution when applying HRTFs in dynamic listening scenarios, the movement of the listener has to be monitored additionally to the modelling of sound source movement [84–86]. The minimum amount of directions and specific measurement points for a sufficiently sparse HRTF set are still current topics of research [87].

HRTFs are listener-specific, i.e., they vary among the listeners [21]. The reasons for that inter-individual variation are usually rooted in listener-specific morphology of the head and ears. For example, the variation in the head width of approximately ±2 cm across the population causes variation in the largest ITD in the range of ±80 μs [88]. **Figure 4** shows HRTF-relevant parts of the human body, where **Figure 4a** shows rough measures of the body and **Figure 4b** shows areas of the pinna responsible for the distinct spectral features in higher frequencies. The width and depth of head and torso have a large effect on HRTFs in the lower frequencies. The inter-individual variation in the pinnae geometry causes variations in HRTFs in frequencies above 4 kHz, with listener-specific differences of up to 20 dB [11]. The inter-individual variation in the HRTFs is rather complex because the pinna is a complex biological structure—small variations in geometry (in the range of millimetres) may cause drastic changes in HRTFs [90] along the vertical planes in high frequencies

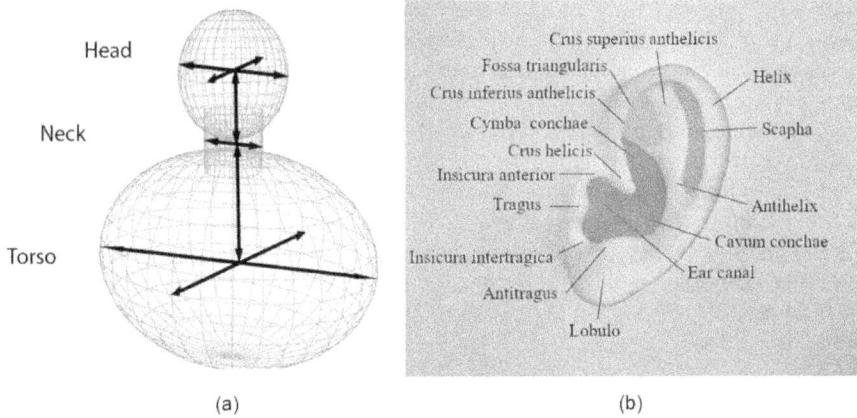

Figure 4.
HRTF-relevant parts of the human body. (a): Head and torso represented with simple shapes based on [57]. The black arrows denote the relevant measures. (b): Pinna and its distinctive regions. In red, green, and blue the concha, fossa triangularis, and scapha, respectively, denote the acoustically relevant areas [48, 56, 89].

[11], see **Figure 2**. However, not all pinna regions affect HRTFs equally [91]. Basically, the convex curvatures of the pinnae contribute to focusing the incoming sound waves towards the entry of the ear canals, comparable to a satellite dish. **Figure 4b** shows the anatomical areas important for localisation of sounds [48, 56, 89, 92, 93]. Currently, the description of the pinna geometry is not a trivial task. Pinnae have been described by means of anthropometric data stored in various data collections, e.g., [67, 69, 88, 94–96]. While the parameters used in these data collections do not seem to completely describe a pinna geometry from scratch, recent efforts aim at parametric pinna models able to generate non-pathological pinna geometries for arbitrary listeners [47, 48]. Such models describe the pinna geometry by means of various control points placed on the surface of a template pinna geometry. **Figure 5** shows two examples of the implementation of such

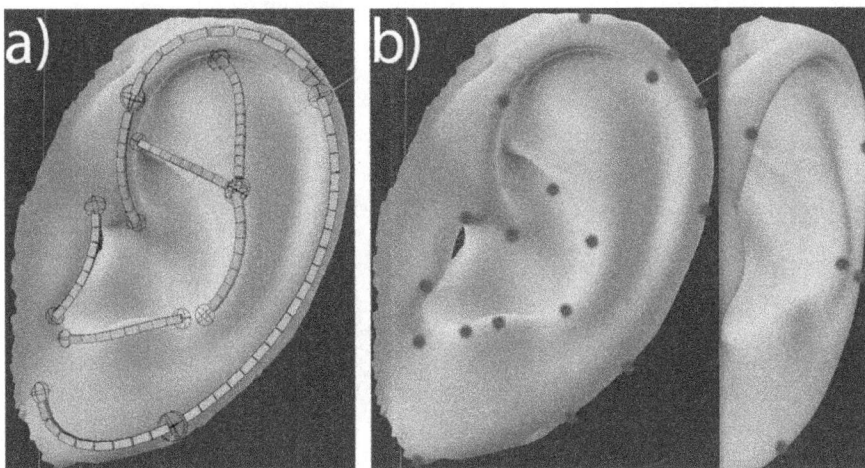

Figure 5.
Examples of parametric pinna models. (a): Model from [47] consisting of Beziér curves (depicted in green), their control points (black spheres at both ends of a curve) and weights (not shown), linked to a template pinna geometry, (b): Model from [48], defined by control points of the pinna relief (green points) linked to proximal mesh vertices.

models. In **Figure 5a**, the pinna geometry is parametrised with the help of Beziér curves, i.e., polynomials within a spatial boundary [47]. **Figure 5b** shows a different approach; here, the parameterisation of the pinna is utilised with control points that move proximal local areas [48]. These parametric pinna models represent a step towards understanding the link between HRTFs and specific anatomical regions of the pinnae, and provide potential to synthesise large datasets of pinnae, e.g., in order to provide data for machine-learning algorithms.

In addition to the geometry, skin and hair may have an impact on HRTFs [97, 98] because of their direction-dependent absorption of the acoustic energy, especially at high frequencies. However, recent studies have shown that hair does not influence the localisation performance, but rather the perception of timbre instead [95, 99–101].

3. Acoustic measurement

The principle of an acoustic HRTF measurement relies on the system identification of the HRTF considered as a linear and time-invariant system. Here, an HRTF describes the propagation path between a microphone and a loudspeaker. Because of the binaural synchronisation, HRTFs are measured simultaneously at the two ears. The measurements are commonly performed for many source positions because of the required high spatial resolution. Recently, the details of the acoustic measurements, including a comprehensive list of HRTF measurement sites has been reviewed [20]. Thus, we only briefly introduce the basics and focus on the most recent advances in the acoustic HRTF measurement.

Typically, two omnidirectional microphones are placed in both ear canals, and the loudspeakers are arranged around the listener, ideally, with the number of loudspeakers corresponding to the number of HRTF positions to be measured. **Figure 6** shows two examples of measurement setups of various complexity: In **Figure 6a**, the listener is located on a turntable and moves within a fixed near-complete circular loudspeaker array. **Figure 6b** shows a similar approach with a near-complete spherical loudspeaker array, and **Figure 6c** shows the placement of a microphone in the ear canal so that it is membrane lines up with the entrance of the

Figure 6.
(a) Example of an HRTF measurement setup with mechanical rotation required. Listener sits on a chair (mounted on a turntable) surrounded by a loudspeaker arc (22 loudspeakers ranging from −30 to 210° in 5°-steps). A head-tracker mounted on the head of the listener tracks head movements, triggering the need for measurement repetition in case of too large movements. (b) Example of more recent HRTF setups. 91 loudspeakers are mounted in a near-complete spherical array reducing the total measurement duration. (c) Example for a microphone placement in an HRTF measurement. Note the closed ear canal and the head-tracker sensor.

ear canal. Actually, it does not matter whether the microphones or loudspeakers are placed in the ear canal—this approach of 'reciprocity' is usually facilitated in numeric HRTF calculations (Section 4.4). However, setups with loudspeakers in the ears [102] lack signal-to-noise ratio (SNR) as the amplitude of the source signal needs to be low enough to not harm the listener, making the setup impractical for experiments. With the microphones in the ears, the most simple setups consist of a single loudspeaker moved around the listener [103]. Unfortunately, such setups lead to a long measurement duration for a dense set of HRTF positions. With the increasing availability of multichannel sound interfaces and adequate electroacoustic equipment, over the decades, the number of actually used loudspeakers increased. Setups with only a single loudspeaker moving around the listener have been replaced by setups with loudspeaker arcs surrounding the listener. In those setups, the listener sits on a turntable and either the listener (e.g., **Figure 6a**) or the loudspeaker arc is rotated [88, 104].

Recent approaches follow one of two different directions; On the one hand, generic and individual HRTFs are measured with a growing number of loudspeakers used in specialised facilities [67, 95]. Some even with such a large amount of loudspeakers that the listener is rotated for a few discrete positions, and postprocessing algorithms interpolate between HRTF directions, e.g., the setup in **Figure 6b**. On the other hand, user-friendly individual HRTF measurement approaches are suggested, showing a trend towards decreasing the complexity of the measurement setup and using widely available equipment. In these approaches, only a single speaker is used and the listener is asked to move the head until a dense setup of HRTF directions can be obtained. These measurements enable simple systems to be used at home [105, 106], in which a head-tracking system records the listener's head movements in real time and adapts the measured spatial HRTF grid. Head-above-torso orientations have to be considered additionally [100], but they reduce the complexity of the measurement setup and enable using widely available equipment, e.g., a commercially available VR headset and one arbitrary loudspeaker, in regular rooms, thus increasing the user-friendliness for setups [105].

Most of those recent approaches consider spatially discrete positions of the listener and/or the loudspeakers. In order to tackle the trade-off between high spatial resolution and long measurement duration, other recent advances have been made towards spatially continuous measurement approaches [107–109]. These approaches enable the measuring of all directions around the listener for a single elevation within less than 4 minutes [110]. Certainly, an advantage of such an approach is the access to the spatially continuous information, which is important especially for frontal HRTF directions. With more and more silent turntables and swivelled chairs, achieving a high SNR is not a big issue. Most recent approaches related to the spatially continuous measurement utilise Kalman filters to acquire system parameters representing HRTFs, and thus speed up the HRTF measurement in a multi-channel setup [111]. Compared to spatially discrete approaches, the spatially continuous method can achieve accuracy within a spectral error of 2 dB [109].

The requirements of the room are not rigorous: In principle, the measurement room does not have to be perfectly anechoic, but it has to fulfil some requirements regarding size and reverberation time. Room modes may exist below 500 Hz as they can be neglected in that frequency range [1]. Acceptable measurement results can be obtained as long as the first room reflection arises after 5 ms such that the measured room impulse responses can be truncated without truncating the HRIRs. Medium and large surfaces, i.e., the mount of the loudspeakers, the loudspeaker arc, the turntable, listener seat, etc., can potentially cause acoustical reflections overlapping with the direct sound path within the first 5 ms of the HRTF. These reflections are usually damped, e.g., by covering the speakers in absorption material. Before the measurement, the listener's head has to be aligned in the measurement setup, adjusting the ears to the interaural axis of the system and the head to the Frankfurt plane. This alignment can be supported by, e.g., a laser system. The orientation and position of the listener's head should be monitored throughout the measurement procedure in order to detect listener's unwanted movements or position drifts. This helps when having to repeat potentially corrupted measurements.

The loudspeakers used for the measurements need to show a fast impulse response decay; fast enough to not interfere with the temporal characteristics of the HRTFs. This can be achieved by using loudspeaker drivers with light membranes, simple electric processing and no acoustic feedback such as a bass-reflex system. The acoustic short-circuit usually limits the lower frequency range of the loudspeakers, and multidriver systems are a common solution to that problem. In order to achieve a spatially compact acoustic source in a multidriver system, it is common to use coaxial loudspeaker drivers with an omnidirectional directivity pattern in HRTF measurement systems [112].

The placement of the microphones can also be an issue. Early setups used an open ear canal where the microphones were positioned close to the eardrum [11]. However, the effect of the ear canal does not seem to be direction-dependent, and its consideration in the measurement introduces technical difficulties and a large measurement variance [19, 113, 114]. Nowadays, the microphones are usually placed at the entrance of the ear canal which is acoustically blocked [11, 20]. Blocking the ear canal can be achieved by using microphones enclosed in earplugs made from foam or silicone or by wrapping the microphone in skin-friendly tape before inserting it. Note that such a measurement captures all directional-dependent features of the acoustic filtering by the outer ear, however, the directional-independent filtering by the ear canal is not captured. All cables from the microphone have to be flexible enough to minimise their effect on the acoustics within the pinna—one way is to lead the cable through the incisura intertragica and secure it with tape on the cheek, see **Figure 6c**.

In general, system identification can be performed with a variety of excitation signals. While previously Golay codes or other broadband signals have been used [115], more recently, the multiple exponential sweep method (MESM) [112] has been established and further improved [116], enabling fast HRTF measurement at high SNRs, reducing the discomfort for the listener. Still because of the imperfections in the electro-acoustic setup, a reference measurement is required to estimate the basis of the measurement without the effect of the listener, i.e., to estimate p_0. It is typically done for each microphone by placing the microphone in the centre of the measurement setup and recording the loudspeaker-microphone impulse response for all loudspeakers. The reference measurement can also be used to control the sound pressure level in order to avoid clipping at the microphones and analogue-digital converters. This can especially happen when each loudspeaker is driven within its linear range, but the overlapping signals from multiple loudspeakers raise the total level to ranges beyond the linear range of the recording system. Additionally, because of the HRTF's resonances, the level during the actual HRTF measurements can be up to 20 dB higher than that during the reference measurement, translating to a requirement for the headroom of at least 30 dB at the reference measurement. The maximum level is not only limited by the equipment; the listener's hearing range also needs to be considered, i.e., the maximum sound pressure level must neither create discomfort for the listener, nor go beyond the levels of safe listening. There is no special requirement for the listener regarding their audible threshold, hearing range or the visual sense. However, a particular measurement equipment or a particular lab could have some restrictions on, e.g., the listener's weight or height.

Figure 7 shows measurement grids of three exemplary setups and one measurement grid of a simulation setup. **Figure 7a** and **b** correspond to the measurement setups in **Figure 6a** and **b**. In these setups, not every loudspeaker plays a stimulus at every position around the listener. An extreme case is a loudspeaker positioned at 90° elevation that needs to be only measured once. **Figure 7c** shows another setup with uniformly distributed measurement points, and **Figure 7d** shows a uniform sampling grid used in numerical calculation experiments [95].

The repeatability of the measurement is an important issue. Within a single laboratory, changes in the room conditions such as temperature and humidity, as well as changes in the setup such as the ageing of the equipment may compromise the repeatability of the HRTF measurement [11, 20]. When comparing the HRTFs measurement across the labs, differences in the setups play also a role. In inter-laboratory and inter-method HRTF measurement comparison obtained for the same artificial head, severe ITD variations of up to 200 μ have been found [63, 64].

Once the HRTFs have been measured for all source positions, post-processing needs to be done before the HRTFs are ready to be used. First, in order to account for acoustic artefacts caused by the measurement room, a frequency-dependent windowing function is usually applied truncating the HRIRs [100, 117, 118]. Second, the measured HRIRs are equalised by the impulse response obtained from the reference measurements, i.e., with the microphone placed at the centre of the coordinate system with the listener absent. This equalisation can be either free-field or diffuse-field. For the free-field equalisation, the reference measurement is required only for the frontal direction (0° azimuth, 0° elevation) [54], whereas for the diffuse-field equalisation, the reference measurement is the root mean square (RMS) impulse response of all directions [75], and the results are commonly denoted as directional transfer functions (DFT) [119]. Third, in most common rooms and even in (semi)anechoic rooms, reflections (or room modes) cause artefacts below 400 Hz, confounding the free-field property of HRTFs. Additionally, most loudspeakers used in the measurement are not able to reproduce low frequencies with sufficient power. Since the listener's anthropometry has a small effect on HRTFs in the low-frequency range, HRTFs can be extrapolated towards lower frequencies with a constant magnitude and linear phase [20, 117]. Further post-processing steps may include spectral smoothing to account for listener position

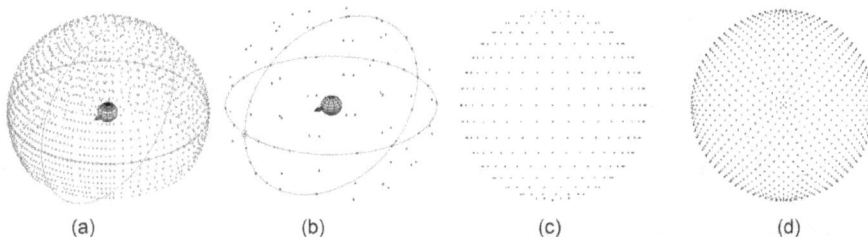

(a)   (b)   (c)   (d)

Figure 7.
*Four examples of spatial HRTF grid resolutions. (a) Almost spherical loudspeaker arc with moving listener, see also **Figure 6a**; the loudspeaker arc consists of 22 loudspeakers and yields 1550 directional measurements. Notice the higher resolution in the front of the listener as opposed to the lower resolution in the lateral regions. (b) Sparse measurement grid in a nearly spherical 91-loudspeaker array, see also **Figure 6b**, yielding 451 directional measurements with 5 different listener positions. (c) Measurement grid with 440 uniformly distributed points, measured with a loudspeaker arc with 37 loudspeakers from the HUTUBS database [95]. (d) Sampling grid for numerical calculations with 1730 directions. (c) and (d) are taken from the HUTUBS database [95].*

inaccuracies [60, 120] or adding a fractional delay to account for temperature changes followed by onset changes of the time signals [100].

The availability of acoustical HRTF measurements was a big step towards personalised binaural audio and virtual reality experience. However, even a fast or continuous measurement method requires the listener to sit still for a few minutes [104, 110, 112] in a specialised lab facility. Recent advances have been made towards both large-scale high-resolution and small-scale at-home easy-to-use solutions, providing HRTF acquisition to a large audience. Still, the imperfections in the electro-acoustic equipment set drawbacks of the acoustic measurement. Here, recent advances in the numeric calculations of the HRTFs can provide an interesting alternative.

4. Numerical calculation of HRTFs

Generally, the calculation of HRTFs simulates the effects of the pinna, head and torso on the sound field at the eardrum. The goal is to numerically obtain the sound pressure at the two ears for a given set of frequencies and spatial positions. There are many methods to simulate wave propagation [121]. When applied to the HRTF calculation, all of the methods require a geometric representation of head and pinnae as input. For an accurate set of HRTFs, an exact 3D representation of the geometry, especially that of the pinnae with all their crests and folds, is of utmost importance [90]. The 3D geometry is represented using a discrete and finite set of elements, further denoted as 'mesh'. A mesh is a representation of the region of interest (ROI), i.e., the object's volume and surface, with the help of simple geometric elements. In most applications, the faces of these elements are assumed to be flat, which in turn explains the preference for triangular faces because they are always flat and therefore have one unique normal vector. This is not always the case for other shapes, e.g., quadrilaterals.

The requirements on the mesh have to consider geometrical as well as acoustical aspects. From the acoustic perspective, a typical rule of thumb for numerical calculation requires the average edge length (AEL) of elements to be at least a sixth of the smallest wavelength [122], which corresponds to an AEL of 3.5 mm for frequencies up to 16 kHz. However, in order to describe the pinna geometry sufficiently accurate, the average edge length (AEL) of the elements in the mesh needs to be around 1 mm, independently of the calculation method [90]. Some numerical calculation algorithms are, in general, more efficient and stable if the geometries are represented locally with elements of similar sizes and as regular as possible, e.g., almost equilateral triangles. To this end, the mesh may undergo a so-called *remeshing* [123], which inserts additional elements and resizes all elements to a similar size. **Figure 8** shows the same pinna in all panels, represented by meshes with increasing AELs from left to right.

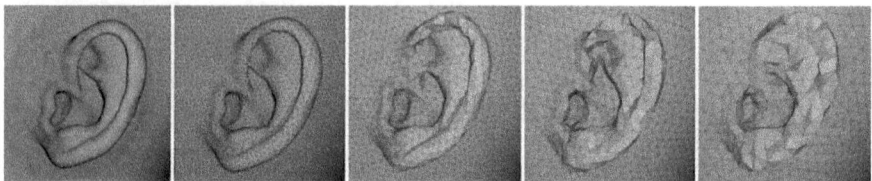

Figure 8.
Pinna meshes represented by various AELs [90]. From left to right: AEL of 1 mm, 2 mm, 3 mm, 4 mm and 5 mm. Note how the representations of the helix and fossa triangularis degrade with increasing AEL.

Interestingly, only the pinna regions contributing to the HRTF (compare **Figure 4b**) require to be accurately represented [56] and the remainder of the geometry can be more roughly modelled. This applies especially to the head, torso and neck, which can be represented by larger elements. These anatomical parts can additionally be approximated by simple geometric shapes, e.g., a sphere for the head, a cylinder for the neck and a rectangular cuboid or an ellipsoid representing the torso [65], see e.g., **Figure 4a**. To emphasise the sophisticated direction dependency of the pinna, **Figure 9** shows the calculated sound pressure distribution over the surface of the pinna. This simulation is calculated by defining one element in the centre of the ear canal as a sound source and evaluating the resulting sound pressure field at the vertices of the rest of the geometry; the procedure is explained thoroughly in Section 4.4.

The geometry can be captured via numerous approaches [124]: a laser scan [125], medical imaging techniques such as magnetic resonance imaging (MRI) [69, 126] and computer tomography (CT) [90], or photogrammetric reconstruction [127]. Laser, MRI and CT scans yield high-resolution meshes offering a small geometric error, but in turn, they need a special equipment. The laser scans are based on line-of-sight propagation and are able to measure short distances with an accuracy of up to 0.01 mm. The downside of line-of-sight propagation is that the manifolds of the pinnae are not easy to capture. In the medical imaging approaches, different downsides arise; acquiring the pinnae geometry via MRI is not a trivial process because they are flattened by the head support. This leads to two separate MRI measurements of each ear. The anatomy is then captured in 'slices' that can be stitched together in the postprocessing rather easily. The CT captures the anatomy in a similar way, but due to the high radiation exposure, such scans are usually not done with human subjects but with (silicone) mouldings of the listener's ear. The overall procedure may take more time than an acoustic HRTF measurement and require the listener to either manufacture a moulding or meeting rather specific criteria for the scanning equipment (e.g., no tattoos, piercings, or implants). As an alternative, recent advances have been made for more widely applicable approaches such as photogrammetry [23, 128]. Photogrammetry is not only non-invasive but also can be done with widely available equipment, e.g., a smartphone or digital

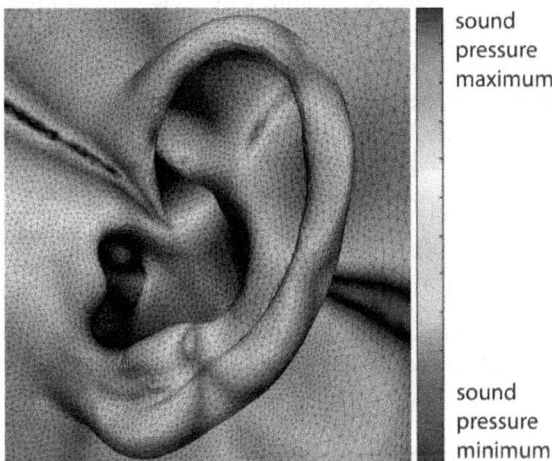

sound
pressure
maximum

sound
pressure
minimum

Figure 9.
Magnitude of the sound pressure calculated for each element of the surface for a 13-kHz sound source placed in the ear canal. Note the high dynamic range containing peaks (red) and notches (blue) in the distribution pattern in the area of the pinna.

camera, without having the listener to travel to a specialised facility. In a nutshell, the photogrammetrical approach works as follows: a set of photographs from different directions is made for each ear [127, 129], the so-called *structure from motion* [130] approach estimates the camera positions by analysing the mutual features across the photographs; a 3D point cloud is constructed; and a 3D mesh is created by connecting the points in the cloud. Note, that currently, manual corrections (e.g., smoothing to reduce noise, filling holes) are still required to reach the high quality of the meshes required for accurate HRTF calculations.

Simulations of acoustics require the information about the acoustic properties of the simulated objects. The HRTFs can be simulated with the 3D geometry represented as fully reflective, i.e., all surfaces having infinite acoustic impedance. With respect to localisation performance, only a small *perceptual* difference was found between acoustically measured and HRTFs calculated for acoustically reflective surfaces [101]. However, the impedance of various regions such as skin and hair may influence the direction-independent HRTF properties and cause changes in the perceived timbre [95, 99, 100].

In order to calculate HRTFs with sufficient spectral accuracy, the number of elements needs to be in the range of several tens of thousands, which might be important for the requirements of the computational power. Such large numerical problems usually require large amount of memory being in the range of Gigabytes. Nevertheless, the calculation time may reach a few days, especially when calculating HRTFs for many frequencies with high-resolution meshes. Note that if the used algorithm calculates HRTFs for each frequency independently, the calculations can be performed in parallel, and computer clusters can be used. This reduces the calculation time to a few hours for HRTFs the full hearing range and a mesh of several tens of thousands of elements.

All the algorithms for numerical HRTF calculation are based on the propagation of sound waves in the free field around a scattering object (also "scatterer"), usually described by the Helmholtz equation

$$\nabla^2 p(\mathbf{x}) + k^2 p(\mathbf{x}) = q(\mathbf{x}), \mathbf{x} \in \Omega_e, \tag{2}$$

where $\nabla^2 = \frac{\partial^2}{\partial x^2} + \frac{\partial^2}{\partial y^2} + \frac{\partial^2}{\partial z^2}$ denotes the Laplace operator in 3D, $p(\mathbf{x})$ denotes the (complex valued) sound pressure at a point $\mathbf{x}$ for a given wavenumber k in the domain Ω_e around the scattering object and $q(\mathbf{x})$ denotes the (complex-valued) contribution of an external sound source at the acoustic field around the object. The wavenumber k is $\frac{2\pi f}{c}$ with f being the frequency and c the speed of sound.

In order to solve the Helmholtz equation for a given scatterer, boundary conditions are necessary. The *Neumann* boundary condition assumes the object to be acoustically hard, and the (scaled) particle velocity at the boundary can be set to zero,

$$\frac{\partial p(\mathbf{x})}{\partial \mathbf{n}} = \nabla p(\mathbf{x}) \cdot \mathbf{n} = 0, \qquad \mathbf{x} \in \Gamma,$$

where $\mathbf{n}$ denotes the normal vector at the surface pointing away from the object. For the boundary condition at infinite distance, the so-called *Sommerfeld* radiation condition can be applied,

$$\left\langle \frac{\mathbf{x}}{\|\mathbf{x}\|}, \nabla p \right\rangle - ikp(\mathbf{x}) = o\left(\frac{1}{\|\mathbf{x}\|}\right), \qquad \|\mathbf{x}\| \to \infty,$$

with $o(.)$ showing that the right side grows much faster than the left side. This ensures that the sound field decays away from the object [131].

For the calculation of HRTFs, the Helmholtz equation can be solved numerically by means of various approaches, which are based on a discretisation of the exterior domain Ω_e around the scatter Γ. Some of these methods solve the Helmholtz equation in the frequency domain, and others solve its counterpart, the wave equation, in the time domain. In general, the formulations and the results in the different domains can be transformed into each other by using, e.g., the Fourier transformation. In the following, we describe the most prominent methods used for HRTF calculations.

4.1 The finite-element method

The finite-element method (FEM) solves the Helmholtz equation, Eq. (2), considering the scattering object or the domain around it as a volume [132]. **Figure 10** shows an example of a finite (domain) volume Ω_e considered in the calculations with the scatterer with surface Γ placed inside that volume. To simulate the acoustic field around that object, the weak form of the Helmholtz equation is used, i.e., the equation is multiplied by a set of known test functions $w(\mathbf{x})$ and integrated over the whole domain, thus transforming the partial differential equation (e.g., the Helmholtz equation) into an integral equation, that can be easier solved numerically:

$$\int_{\Omega_e} (\nabla^2 p(\mathbf{x}) + k^2 p(\mathbf{x}))w(\mathbf{x})d\mathbf{x} = \int_{\Omega_e} q(\mathbf{x})w(\mathbf{x})d\mathbf{x}. \tag{3}$$

Secondly, the unknown pressure $p(\mathbf{x})$ is approximated by a linear combination

$$p(\mathbf{x}) = \sum_{j=1}^{N} p_j \phi_j(\mathbf{x}) \tag{4}$$

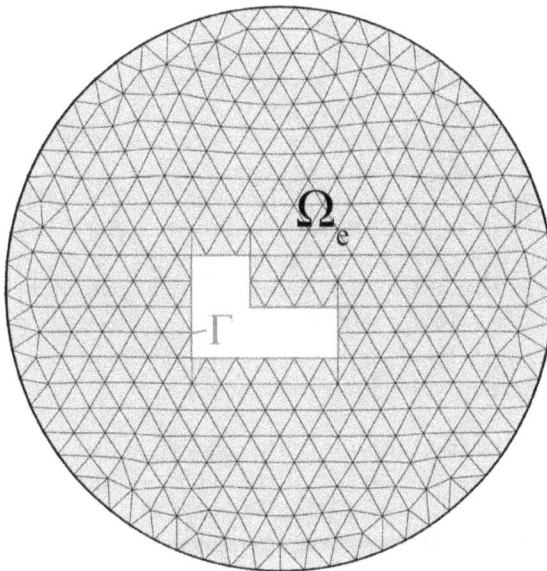

Figure 10.
2D representation of meshes used in FEM. The elements are uniformly distributed and fitted to the boundary of the domain Ω_e.

of so-called ansatz functions $\phi_j(\mathbf{x}), j = 1, \ldots, N$. These ansatz functions, or element shape functions, help at interpolating between the discrete solutions for each point of the mesh. They are, in general, simple (real) polynomials defined locally on the elements of the mesh, e.g., having the value of 1 at their own coordinates and zero for other points of the mesh. Recent advances have been made towards adaptively finding higher-order polynomials and thus drastically reducing the computational effort [133, 134]. In theory, Eq. (3) should be fulfilled for all possible test functions $w(\mathbf{x})$, in practice, however, often the ansatz functions are also used as test functions, i.e., $w_i(\mathbf{x}) = \phi_i(\mathbf{x})$. With this choice, Eq. (3) can be transformed into a linear system of equations

$$\mathbf{Kp} = \mathbf{g}, \tag{5}$$

where

$$\mathbf{K}_{ij} = \int_{\Omega_e} \frac{d\phi_i}{d\mathbf{x}} \frac{d\phi_j}{d\mathbf{x}} - k^2 \phi_i \phi_j d\mathbf{x}, \qquad \mathbf{g}_i = \int_{\Omega_e} q(\mathbf{x}) \phi_i(\mathbf{x}) d\mathbf{x},$$

and $\mathbf{p}$ is the vector containing the unknown coefficients p_i of the representation Eq. (4).

In general, the unknown coefficients p_i represent the complex sound pressure at a given node $\mathbf{x}_i$ of the mesh. The integrals involved are solved using numerical methods [135].

When calculating HRTFs, the space around the scatterer is assumed to be continuous and infinite; in practice, this space has to be discretised and truncated to a finite domain by inserting a virtual boundary. When applied to the calculation of HRTFs, a virtual boundary of the (now finite) domain Ω_e needs to be defined and conditions have to be set to avoid any reflections from this boundary, thus keeping in line with anechoic or free-field conditions. There are several methods to do so, with the so-called perfectly matched layers (PMLs) being the most popular among the reviewed HRTF calculation approaches. The PML is created by inserting an artificial boundary inside Ω_e, e.g., a sphere with sufficiently large radius, and artificially damped equations are used to represent a solution that can then be numerically calculated, fulfilling the Sommerfeld radiation condition. Recent advances have been made to define PMLs automatically by extruding the boundary layer of the mesh and obtaining the geometric parameters during the extrusion step [136].

The FEM has been widely used in HRTF calculations [137–141] and yields similar results to acoustical HRTF measurements with spectral magnitude errors of approximately 1 dB [137, 141]. The downside, however, is the need to model 3D volumes around the head, resulting in models of a high number of elements, having a strong impact on the calculation duration.

4.2 The finite-difference time-domain method

A similar approach as the FEM can also be followed in the time domain. By using a short sound burst in the time domain as an input signal, the HRTFs within a wide frequency range can be calculated at once. This approach is called the finite-difference time-domain (FDTD) method [142] and can be derived by solving the wave equation in the time domain

$$\nabla^2 \breve{p}(\mathbf{x}, t) - \frac{\partial^2 \breve{p}(\mathbf{x}, t)}{\partial t^2} = \breve{q}(\mathbf{x}, t),$$

where $\breve{p}$ and $\breve{q}$ denote sound pressure fields in the time domain. The PML is applied to create the boundary conditions for outgoing sound waves. The evaluation grid is sampled evenly in cells across the domain with grid spacing h, considering discrete time steps m. A key parameter for numerical stability of the FDTD is the Courant number

$$\beta = \frac{cm}{h},$$

defining the number of cells the sound propagates per time step. Typically, in order to obtain stable HRTF calculations, the Courant number is $\beta = 1/\sqrt{3}$.

Figure 11 shows a 2D representation of a mesh used in the FDTD method. Note that because the mesh needs to consist of evenly spaced elements, most of the objects cannot be represented accurately and a sampling error is introduced at the boundary surface Γ of the object. Additionally, as derivatives of functions are approximated by finite differences, the arithmetic operations are valid for infinite resolution, but when calculating on physical computers, the precision depends on the number format used and the gridsize, introducing errors in the results [143]. Refining the grid is a potential solution to the sampling error for staircase approximations [144, 145], and when framing this problem to HRTF calculations, spectral magnitude errors of 1 dB up to 8 kHz and 2 dB up to 18 kHz can be achieved [146, 147], suggesting only negligible increase in localisation errors when listening to HRTFs calculated by the FDTD method.

Because of the additional sampling errors for irregular domains, recent advances have been made towards using quasi-cartesian grids [148], dynamically choosing grid resolutions [149], or towards the finite-volume method (FVTD), which is based on energy conservation and dissipation of the system as a whole and uses the integral formulation of the FDTD [150]. One solution approach there is to adaptively sample the grid at the boundary and introduce unstructured or fitted cells [151, 152]. A thorough comparison between FEM, FDTD and FVTD methods is available in [153].

In fact, the FDTD method has been widely applied to HRTF calculations [145, 146, 154, 155], and it certainly offers the advantage of calculating broadband

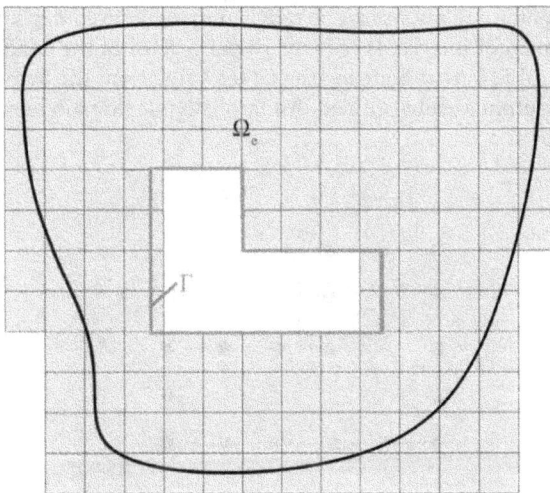

Figure 11.
2D representation of meshes used in the FDTD method. Note that in this representation, the object surface Γ does not line up exactly with the sampling grid.

HRTFs while not introducing additional computational cost when multiple inputs or outputs are used. However, because of the complex geometry of the pinnae, a submillimetre sampling grid is required, resulting in the need for a delicate preprocessing.

4.3 The boundary-element method

The boundary element method (BEM) is based on a special set of test functions in the weak formulation of the Helmholtz equation Eq. (3), namely the Green's function

$$G(\mathbf{x}, \mathbf{y}) = \frac{e^{ikr}}{4\pi r},$$

where $r = \|\mathbf{x} - \mathbf{y}\|$, and $\mathbf{x}$ and $\mathbf{y}$ are two points in space. By using this function, it is possible to reduce the weak form of the Helmholtz equation to an integral equation, i.e., the boundary integral equation (BIE), that only involves integrals over the surface Γ of the object, and *not* the volume Ω_e. Assuming that the external sound source as a point source at $\mathbf{x}^*$, and an acoustically reflecting (= sound hard, $\nabla p(\mathbf{y}) \cdot \mathbf{n} = 0, \mathbf{y} \in \Gamma$) surface, the sound field at a point $\mathbf{x}$ on a smooth part of that surface Γ is given by:

$$\frac{1}{2}p(\mathbf{x}) - \int_\Gamma H(\mathbf{x}, \mathbf{y})p(\mathbf{y})d\mathbf{y} = G(\mathbf{x}^*, \mathbf{x})p_0, \qquad (6)$$

where $H(\mathbf{x}, \mathbf{y}) = \frac{\partial G}{\partial \mathbf{n}_y}(\mathbf{x}, \mathbf{y})$ is obtained by the derivative of the Green's function at a point $\mathbf{y}$ in the direction of vector $\mathbf{n}$ normal to the surface at this position, and p_0 denotes the strength of the sound source.

In comparison with the other two methods, the BEM has the advantage that only the *surface* of the object such as the head and the pinnae needs to be discretised, whereas in FEM and the FDTD method also a discretisation of the *volume* surrounding the head has to be considered, see **Figures 10–12**. Thus, in the boundary element method, all calculations can be reduced to a manifold described in 2D, in our case, the domain of interest is reduced to the surface of the head. A second advantage of the BEM is that by using the Green's function, the Sommerfeld radiation condition is automatically fulfilled. Additionally, no domain boundary has to be

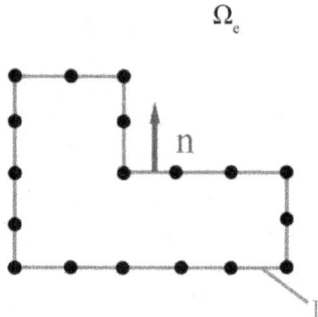

Figure 12.
2D representation of a BEM mesh. Note that only the boundary of the surface Γ is considered and the domain volume Ω_e does not have to be sampled.

introduced, such as the PML. This renders the BEM an attractive method for calculating sound propagation in infinite domains, i.e., in free-field, as is the assumption when calculating HRTFs [156].

In order to solve a BEM problem, the BIE is discretized and solved by using methods such as the Galerkin, collocation or Nyström [157–159], all with the common goal of yielding a linear system of equations.

For the Galerkin method, the unknown pressure is approximated by a linear combination of ansatz functions as in Eq. (4). The BIE is again multiplied with a set of test functions (similar to the test functions $\phi_i(\mathbf{x})$ used in FEM) and integrated with respect to $\mathbf{x}$ yielding a linear system of equations as in Eq. (5), where

$$\mathbf{K_{ij}} = \int_\Gamma \frac{\phi_i(\mathbf{x})\phi_j(\mathbf{x})}{2} d\mathbf{x} - \iint_\Gamma \phi_i(\mathbf{x})H(\mathbf{x},\mathbf{y})\phi_j(\mathbf{y})d\mathbf{y}d\mathbf{x},$$

and

$$\mathbf{g}_i = \int_\Gamma G(\mathbf{x}_0 - \mathbf{x})\phi_i(\mathbf{x})d\mathbf{x}.$$

Another commonly used approach especially used in engineering is collocation with constant elements, i.e., the sound field is assumed to be constant on each element of the mesh, and the BIE is solved at the midpoints $\mathbf{x}_i$ of each element (the set of all $\mathbf{x}_i$ are called collocation nodes) yielding a linear system of equations as in Eq. (5), where

$$\mathbf{K_{ij}} = \frac{1}{2}\delta_{ij} - \int_{\Gamma_j} H(\mathbf{x}_i,\mathbf{y})d\Gamma, \qquad \mathbf{g}_i = G(\mathbf{x}^*,\mathbf{x}_i)p_0.$$

$\mathbf{p}_i = p(\mathbf{x}_i)$ and with $\mathbf{x}^*$ being the position of the sound source outside the head. The integrals over each element are numerically solved using appropriate quadrature formulas (weighted sum of function values) and

$$\delta_{ij} = \begin{cases} 1 & \text{for } i = j \\ 0 & \text{for } i \neq j \end{cases}$$

The BIE is solved for a given set of frequencies and the solutions $\mathbf{p}_i$ at the collocation nodes are then used to derive the HRTFs. Note that the collocation method can be interpreted as the Galerkin method utilising the delta functionals $\delta(\mathbf{x}_i - \mathbf{x})$ as test functions. A thorough comparison between Galerkin and collocation approaches can be found in [160].

The discretisation of just the surface introduces additional challenges. First, the Green's function becomes singular at the boundary where $\|\mathbf{x} - \mathbf{y}\| = 0$ and special quadrature formulas need to be used close to these singularities [161, 162]; and second, the system matrix $\mathbf{K}$, although small, is usually densely populated, which poses a challenge for computer memory and the efficiency of the linear solver used. When considering HRTF calculations for frequencies up to 20 kHz, high-resolution meshes are required and the corresponding linear systems may contain up to 100,000 unknowns.

In order to efficiently deal with such large systems, the BEM can be coupled with methods speeding up matrix–vector multiplications, such as the fast-multipole method (FMM) [163] or $\mathcal{H}$-matrices [164] (so-called 'hierarchical' matrices). These methods have enabled an efficient and feasible calculation of HRTFs [165]. In a nutshell, these methods aim at providing a method for an efficient and fast matrix–vector multiplication and are based on two steps. First, the elements of the mesh are grouped into clusters of approximately the same size with cluster midpoints z_i. Second, for two clusters C_1 and C_2, that are sufficiently apart from each other, a separable approximation of the Green's function

$$G(\mathbf{x}, \mathbf{y}) \approx G_1(\mathbf{x} - \mathbf{z_1})M(\mathbf{z_1}, \mathbf{z_2})G_2(\mathbf{y} - \mathbf{z_2})$$

is found. This approximation has two advantages: the local expansions G_1 and G_2 have to be made only once for each cluster, and the interaction between the elements of different clusters is reduced to a single interaction of the cluster midpoints. The resulting linear system of equations is then solved using an iterative equation solver [166].

Although the Helmholtz equation for external problems has a unique solution at all frequencies, the BIE has uniqueness problems at certain critical frequencies [159, 167]. Thus, to avoid numerical problems, the BEM needs to be stabilised at these frequencies, e.g., by using the Burton-Miller method [167]. BEM has been widely used to calculate HRTFs [165, 168–171] analysing the process from various perspectives. When applied to an accurate and high-resolution representation of the pinna geometry, BEM can yield similar results to the acoustic HRTF measurements by means of sound localisation performance [101, 172].

4.4 Reciprocity

In principle, in order to calculate an HRTF set, the Helmholtz equation needs to be solved for every source position $\mathbf{x}_j^*$ separately, leading in up to several thousand right-hand sides in Eq. (5). Solving that many equations cannot be done quickly even with the help of iterative solvers. On the other hand, the HRTF calculation for the second ear is quite simple because the solution obtained from the solver is already available for every element of the surface, i.e., at the element representing the ear canal of the second ear. The approach of reciprocity can help to significantly speed up the calculations by swapping the role of the many source positions with the two elements representing the ear canals, requiring us to solve Eq. (5) only twice, i.e., once for each of the ears.

Helmholtz' reciprocity theorem states that switching source and receiver positions do not affect the observed sound pressure. When applied to HRTF calculations, virtual loudspeakers are placed in the entrance of the ear canal (replacing the virtual microphones) and the many simulated sound sources are represented by many virtual microphones (replacing the many virtual loudspeakers around the listener). By doing so, the computationally expensive part of the BEM, i.e., solving a linear system of equations to calculate the sound pressure at the surface, needs to be done only twice, namely once for each ear. Subsequently, the sound pressure at positions around the head can be calculated fairly easy and efficiently.

In more detail, assume that a point source with strength p_0 at the position $\mathbf{x}_j^*$ causes a mean sound pressure of $\bar{p}$ over a small domain with area A_0 at the entrance of the ear canal. If the domain is sufficiently small, the mean sound pressure is an accurate representation of the actual sound pressure in this domain. By applying the reciprocity, we introduce a reciprocal sound source at the entrance of the ear canal which introduces a velocity v_0 and then calculate the sound pressure $p\left(\mathbf{x}_j^*\right)$ at the actual sound-source position around the listener. The pressures $p\left(\mathbf{x}_j^*\right)$ and $\bar{p}$ are linked by

$$\bar{p} = \frac{p_0 p\left(\mathbf{x}_j^*\right)}{A_0 v_0}.$$

The reciprocal sound source can be modelled by vibrating elements Γ_{vib}, i.e., elements with an additional velocity boundary condition

$$v(\mathbf{y}) = \frac{1}{i\omega\rho}\frac{\partial p}{\partial \mathbf{n}} = v_0, \qquad \mathbf{y} \in \Gamma_{\mathrm{vib}}$$

where $\omega = 2\pi f_\ell$, and ρ is the density of air. Note that v_0 can be an arbitrary positive number because when calculating HRTFs [see for example Eq. (1)], the pressure is normalised by the reference pressure p_0, thus cancelling v_0. With this additional boundary condition, first, BEM can be used to calculate the sound field at the surface Γ, and then, Green's representation is applied to calculate the pressure at all positions of the actual sound sources $\mathbf{x}_j^*$,

$$p\left(\mathbf{x}_j^*\right) - \int_\Gamma H\left(\mathbf{x}_j^*, \mathbf{y}\right)p(\mathbf{y})d\mathbf{y} + i\omega\rho \int_\Gamma G\left(\mathbf{x}_j^*, \mathbf{y}\right)v(\mathbf{y})d\mathbf{y} = 0.$$

Note that this equation is calculated after a discretisation, and because $p(\mathbf{y})$ at the surface is known from the BEM solution, the calculation of the sound pressure around the head $p\left(\mathbf{x}_j^*\right)$ is a simple matrix multiplication.

Reciprocity, combined with FMM-coupled BEM has been applied to calculate HRTFs, enabling calculations for a large spatial HRTF set within a few hours even on a standard desktop computer [172].

5. Other issues related to HRTF acquisition

Over decades, HRTFs have been collected and stored in databases. Such databases are important for educational aspects, training of neural network algorithms [34, 37] and further research [23, 25–28, 173]. While in the early HRTF research days, HRTFs have been stored by each lab in a different format, since 2015, the spatially oriented format for acoustics (SOFA) is available to store HRTFs in a flexible but well-described way facilitating an easy exchange between the labs and applications. SOFA is a standard of the Audio Engineering Society under the name AES69. SOFA provides a uniform description of spatially oriented acoustic data such as HRTFs, spatial room impulse responses, and directivities [15].

When it comes to anthropometric data, unfortunately, there is currently no common format to specify and exchange anthropometric data. This is partially because currently, it is not known, which data are important. Some laboratories use the CIPIC parameters [88], some have extended them [174], and others have created whole new sets of parameters [128, 175]. An overview of currently used anthropometric parameters can be found in [176]. The development of parametric pinna models may shed light on the relevance of parameters needed to be stored in the future. The listener's geometry can also be stored in non-parametric representations such as meshes and point clouds of listener's ears and head. To this end, typical 3D dataset formats are used, e.g., OBJ, PLY or STL. These formats are widely used in computer graphics and thus easily accessible by many corresponding applications. A large collection of HRTF databases stored in SOFA, with some of them combined with meshes stored in OBJ, PLY and STL files is available at the SOFA website.[1]

When HRTFs are obtained, there is strong demand to evaluate their quality. This is especially interesting when comparing the results from numerical HRTF calculations. The evaluations can be performed at various levels: geometrical, acoustical and perceptive. The evaluation at the geometric level can be done by comparing the deviation between two meshes of the pinna and representing the deviation as the Hausdorff distance [177]. The evaluation at the acoustic level can be done by calculating the spectral distortion

$$SD = \sqrt{\frac{1}{N} \sum_{i=1}^{N} \left(20 \log \frac{\widehat{HRTF}(\mathbf{x}^*, f_i)}{HRTF(\mathbf{x}^*, f_i)} \right)^2}, \tag{7}$$

where $\widehat{HRTF}(\mathbf{x}^*, f_i)$ denotes the calculated and $HRTF(\mathbf{x}^*, f_i)$ the measured one, summarised over N discrete frequencies. The evaluation at the perceptual level can be simulated by means of auditory modelling [50] or direct performance of localisation experiment [50, 90, 147]. Especially the evaluation of localisation errors in the median plane can be relevant because the sound localisation in the median plane is directly related to the quality of monaural spectral features in the HRTFs [46, 178]. A calculated HRTF set yielding similar perceptual results as the natural listener's HRTFs can be described as *perceptually valid*.

6. Conclusions

With a specialised measurement setup, acoustic HRTF measurements can be done within a few minutes. Still, such setups are expensive and require the listener to sit or stand still for the whole measurement duration. The requirement of specialised components has been limiting the popularity of the acoustic methods. Recent advances, however, have been made by integrating head-movement tracking in systems to be used at home, especially since the commercialisation of VR headsets. These advances provide an easy-to-use measurement setup, but still need investigation on how many and which measurement positions are crucial to acquire a sufficient measurement grid for perceptually valid HRTFs.

With the availability of numerical HRTF calculations, the acquisition of personalised HRTFs has undergone significant advances. While the acoustic HRTF measurement still remains the reference acquisition method, numerical HRTF calculation paves the road towards personalised HRTFs available for a wide audience. The most widely used approaches, FEM, FDTD, BEM and BEM coupled with the FMM, when applied under optimal conditions, can yield acoustically and perceptually valid results.

Machine learning and neural networks gain increasing popularity and, in the future, may even further push the usability of numerical HRTF calculations. For example, neural networks might be able to support the photogrammetric mesh acquisition or even estimate the HRTFs directly from listener-specific anthropometric data such as photographs. Further improvements in terms of efficiency, accuracy and precision are still ongoing subject of research.

Despite the clear definition when it comes to storing an HRTF data set by means of SOFA, a similar definition for the description of anthropometric data is still not

---

[1] https://www.sofaconventions.org/mediawiki/index.php/Files

available. This might be rooted in our poor understanding of the importance of parts of the pinna and its contribution to the HRTF. Here, a clear goal is to better understand the anthropometry and its relation with HRTFs. All this future work heads into the direction of expanding the access to personalised HRTFs enabling their availability for everyone.

Acknowledgements

This work was supported by the Austrian Research Promotion Agency (FFG, project 'softpinna' 871263) and the European Union (EU, project 'SONICOM' 101017743, RIA action of Horizon 2020). We thank Harald Ziegelwanger for visualising the sound pressure in Figure 9.

Conflict of interest

The authors declare no conflict of interest.

Author details

Katharina Pollack*, Wolfgang Kreuzer and Piotr Majdak
Acoustics Research Institute, Austrian Academy of Sciences, Vienna, Austria

*Address all correspondence to: katharina.pollack@oeaw.ac.at

IntechOpen

References

[1] Algazi VR, Avendano C, Duda RO. Elevation localization and head-related transfer function analysis at low frequencies. The Journal of the Acoustical Society of America. 2001; **109**(3):1110-1122. DOI: 10.1121/1.1349185

[2] Batteau DW. The role of the pinna in human localization. Proceedings of the Royal Society of London Series B. Biological Sciences. 1967;**168**(1011): 158-180. DOI: 10.1098/rspb.1967.0058

[3] Baumgartner R, Reed DK, Tóth B, Best V, Majdak P, Colburn HS, et al. Asymmetries in behavioral and neural responses to spectral cues demonstrate the generality of auditory looming bias. Proceedings of the National Academy of Sciences. 2017;**114**(36):9743-9748, ISSN: 0027-8424, 1091-6490. DOI: 10.1073/pnas.1703247114

[4] Fisher HG, Freedman SJ. The role of the pinna in auditory localization. Journal of Auditory Research. 1968; **168**(1011):158-180

[5] Hebrank J, Wright D. Spectral cues used in the localization of sound sources on the median plane. The Journal of the Acoustical Society of America. 1974; **56**(6):1829-1834. DOI: 10.1121/1.1903520

[6] Musicant AD, Butler RA. The influence of pinnae-based spectral cues on sound localization. The Journal of the Acoustical Society of America. 1984; 75(4):1195-1200. DOI: 10.1121/1.390770

[7] Majdak P, Baumgartner R, Jenny C. Formation of three-dimensional auditory space. In: Blauert J, Braasch J, editors. The Technology of Binaural Understanding, Modern Acoustics and Signal Processing. Cham, ISBN: 978-3-030-00386-9: Springer International Publishing; 2020. pp. 115-149. DOI: 10.1007/978-3-030-00386-9_5

[8] Majdak P, Baumgartner R, Laback B. Acoustic and non-acoustic factors in modeling listener-specific performance of sagittal-plane sound localization. Frontiers in Psychology. 2014;**5**:319. DOI: 10.3389/fpsyg.2014.00319

[9] Seeber BU, Fastl H. Subjective selection of non-individual head-related transfer functions. In: Proceedings of the International Conference on Auditory Display. Atlanta, Georgia: Georgia Institute of Technology; 2003. pp. 259-262

[10] Wenzel EM, Arruda M, Kistler DJ, Wightman FL. Localization using nonindividualized head-related transfer functions. The Journal of the Acoustical Society of America. 1993;**94**(1):111-123. DOI: 10.1121/1.407089

[11] Møller H, Sørensen MF, Hammershøi D, Jensen CB. Head-related transfer functions of human subjects. Journal of the Audio Engineering Society. 1995;**43**:300-321

[12] Macpherson EA, Middlebrooks JC. Listener weighting of cues for lateral angle: The duplex theory of sound localization revisited. The Journal of the Acoustical Society of America. 2002;**111** (5 Pt 1):2219-2236. DOI: 10.1121/1.1471898

[13] Reijniers J, Vanderelst D, Jin C, Carlile S, Peremans H. An ideal-observer model of human sound localization. Biological Cybernetics. 2014;**108**(2):169-181, ISSN: 0340-1200. DOI: 10.1007/s00422-014-0588-4

[14] Majdak P, Goupell MJ, Laback B. 3-d localization of virtual sound sources: Effects of visual environment, pointing method, and training. Attention, Perception, & Psychophysics. 2010; 72(2):454-469. DOI: 10.3758/APP.72.2.454

[15] Majdak P, Carpentier T, Nicol R, Roginska A, Suzuki Y, Watanabe K, et al. Spatially oriented format for acoustics: A data exchange format representing head-related transfer functions. In: Proceedings of the 134th Convention of the Audio Engineering Society (AES), Page Convention Paper 8880. Roma, Italy: Audio Engineering Society; 2013

[16] Majdak P, Hollomey C, Baumgartner R. The auditory modeling toolbox. In: The Technology of Binaural Listening. Berlin, Heidelberg: Springer; 2021. pp. 33-56

[17] Søndergaard P, Majdak P. The auditory modeling toolbox. In: Blauert J, editor. The Technology of Binaural Listening. Berlin-Heidelberg, Germany: Springer; 2013. pp. 33-56. DOI: 10.1007/978-3-642-37762-4_2

[18] Guezenoc C, Seguier R. HRTF individualization: A survey. In Audio Engineering Society convention 145, page Convention Paper 10129. New York, New York, United States: Audio Engineering Society; 2018

[19] Hammershøi D, Møller H. Sound transmission to and within the human ear canal. The Journal of the Acoustical Society of America. 1996;**100**(1): 408-427. DOI: 10.1121/1.415856

[20] Li S, Peissig J. Measurement of head-related transfer functions: A review. Applied Sciences. 2020;**10**(14): 5014. DOI: 10.3390/app101450140 Number: 14 Publisher: Multidisciplinary Digital Publishing Institute

[21] Middlebrooks JC. Individual differences in external-ear transfer functions reduced by scaling in frequency. The Journal of the Acoustical Society of America. 1999;**106**(3): 1480-1492. DOI: 10.1121/1.427176

[22] Iida K, Aizaki T, Kikuchi T. Toolkit for individualization of head-related transfer functions using parametric notch-peak model. Applied Acoustics. 2022;**189**:108610. DOI: 10.1016/j.apacoust.2021.108610

[23] Torres-Gallegos EA, Orduna-Bustamante F, Arámbula-Cosío F. Personalization of head-related transfer functions (HRTF) based on automatic photo-anthropometry and inference from a database. Applied Acoustics. 2015;**97**:84-95. DOI: 10.1016/j.apacoust.2015.04.009

[24] Guezenoc C, Seguier R. A wide dataset of ear shapes and pinna-related transfer functions generated by random ear drawings. The Journal of the Acoustical Society of America. 2020; **147**(6):4087-4096. DOI: 10.1121/10.0001461

[25] Jin CT, Zolfaghari R, Long X, Sebastian A, Hossain S, Glaunés J, et al. Considerations regarding individualization of head-related transfer functions. In: 2018 IEEE International Conference on Acoustics, Speech and Signal Processing (ICASSP). Calgary, AB, Canada: IEEE; 2018. pp. 6787-6791. DOI: 10.1109/ICASSP.2018.8462613

[26] Lu D, Zeng X, Guo X, Wang H. Personalization of head-related transfer function based on sparse principle component analysis and sparse representation of 3d anthropometric parameters. Australia: Acoustics; 2019. pp. 1-10. DOI: 10.1007/s40857-019-00169-y

[27] Tommasini FC, Ramos OA, Hüg MX, Bermejo F. Usage of spectral distortion for objective evaluation of personalized hrtf in the median plane. International Journal of Acoustics & Vibration. 2015;**20**(2):81-89

[28] Zhang M, Ge Z, Liu T, Wu X, Qu T. Modeling of individual HRTFs based on spatial principal component analysis. IEEE/ACM Transactions on Audio,

Speech, and Language Processing. 2020;
28:785-797. DOI: 10.1109/
TASLP.2020.2967539

[29] Zhang M, Kennedy R,
Abhayapala T, Zhang W. Statistical
method to identify key anthropometric
parameters in HRTF individualization.
In: 2011 Joint Workshop on Hands-free
Speech Communication and
Microphone Arrays. Edinburgh,
Scotland: IEEE; 2011. pp. 213-218. DOI:
10.1109/HSCMA.2011.5942401

[30] Hu H, Zhou L, Zhang J, Ma H,
Wu Z. Head related transfer function
personalization based on multiple
regression analysis. In: 2006
International Conference on
Computational Intelligence and
Security. Vol. 2. Guangzhou, China:
IEEE; 2006. pp. 1829-1832. DOI:
10.1109/ICCIAS.2006.295380

[31] Huang Q, Zhuang Q. HRIR
personalisation using support vector
regression in independent feature space.
Electronics Letters. 2009;**45**(19):
1002-1003

[32] Zolfaghari R, Epain N, Jin CT,
Glaunes J, Tew A. Large deformation
diffeomorphic metric mapping and fast-
multipole boundary element method
provide new insights for binaural
acoustics. In: 2014 IEEE International
Conference on Acoustics, Speech and
Signal Processing (ICASSP). London:
IEEE; 2014. pp. 2863-2867. DOI:
10.1109/ICASSP.2014.6854123

[33] Grijalva F, Martini LC, Florencio D,
Goldenstein S. Interpolation of head-
related transfer functions using
manifold learning. IEEE Signal
Processing Letters. 2017;**24**(2):221-225.
DOI: 10.1109/LSP.2017.2648794

[34] Gebru ID, Marković D, Richard A,
Krenn S, Butler GA, De la Torre F, et al.
Implicit HRTF modeling using temporal
convolutional networks. In: ICASSP
2021-2021 IEEE International

Conference on Acoustics, Speech and
Signal Processing (ICASSP). Singapore:
IEEE; 2021. pp. 3385-3389. DOI:
10.1109/ICASSP39728.2021.9414750

[35] Grijalva F, Martini L, Goldenstein S,
Florencio D. Anthropometric-based
customization of head-related transfer
functions using isomap in the horizontal
plane. In: 2014 IEEE International
Conference on Acoustics, Speech and
Signal Processing (ICASSP). USA:
IEEE; 2014. pp. 4473-4477. DOI:
10.1109/ICASSP.2014.6854448

[36] Hu H, Zhou L, Ma H, Wu Z. HRTF
personalization based on artificial neural
network in individual virtual auditory
space. Applied Acoustics. 2008;**69**(2):
163-172. DOI: 10.1016/j.apacoust.2007.
05.007

[37] Lee GW, Lee JH, Kim SJ, Kim HK.
Directional audio rendering using a
neural network based personalized
HRTF. In INTERSPEECH, Brno, Czech
Republic. pp. 2364–2365

[38] Li L, Huang Q. HRTF personalization
modeling based on RBF neural network.
In: 2013 IEEE International Conference on
Acoustics, Speech and Signal Processing.
Vancouver, Canada: IEEE; 2013.
pp. 3707-3710. DOI: 10.1109/
ICASSP.2013.6638350

[39] Miccini R, Spagnol S. A hybrid
approach to structural modeling of
individualized HRTFs. In: 2021 IEEE
Conference on Virtual Reality and 3D
User Interfaces Abstracts and
Workshops (VRW). Lisbon, Portugal:
IEEE; 2021. pp. 80-85. DOI: 10.1109/
VRW52623.2021.00022

[40] Shu-Nung Y, Collins T, Liang C.
Head-related transfer function selection
using neural networks. Archives of
Acoustics. 2017;**42**(3):365-373. DOI:
10.1515/aoa-2017-0038

[41] Zhou Y, Jiang H, Ithapu VK. On the
predictability of HRTFs from ear shapes

using deep networks. In: ICASSP 2021-2021 IEEE International Conference on Acoustics, Speech and Signal Processing (ICASSP). London: IEEE; 2021. pp. 441-445. DOI: 10.1109/ICASSP39728.2021.9414042

[42] Bilinski P, Ahrens J, Thomas MR, Tashev IJ, Platt JC. HRTF magnitude synthesis via sparse representation of anthropometric features. In: 2014 IEEE International Conference on Acoustics, Speech and Signal Processing (ICASSP). London: IEEE; 2014. pp. 4468-4472. DOI: 10.1109/ICASSP.2014.6854447

[43] Ghorbal S, Auclair T, Soladie C, Seguier R. Pinna morphological parameters influencing HRTF sets. In: Proceedings of the 20th International Conference on Digital Audio Effects (DAFx-17). Edinburgh: University of Edinburgh; 2017. pp. 353-359

[44] Mokhtari P, Takemoto H, Nishimura R, Kato H. Vertical normal modes of human ears: Individual variation and frequency estimation from pinna anthropometry. The Journal of the Acoustical Society of America. 2016;140(2):814-831. DOI: 10.1121/1.4960481

[45] Onofrei MG, Miccini R, Unnthorsson R, Serafin S, Spagnol S. 3d ear shape as an estimator of HRTF notch frequency. In: 17th Sound and Music Computing Conference. Torino: Sound and Music Computing Network; 2020. pp. 131-137. DOI: 10.5281/zenodo.3898720

[46] Spagnol S, Geronazzo M, Avanzini F. On the relation between pinna reflection patterns and head-related transfer function features. IEEE Transactions on Audio, Speech, and Language Processing. 2012;21(3):508-519. DOI: 10.1109/TASL.2012.2227730

[47] Pollack K, Majdak P, Furtado H. A parametric pinna model for the

calculations of head-related transfer functions. In: Proceedings of Forum Acusticum. Lyon. 2020. pp. 1357-1360. DOI: 10.48465/fa.2020.02800

[48] Stitt P, Katz BFG. Sensitivity analysis of pinna morphology on head-related transfer functions simulated via a parametric pinna model. The Journal of the Acoustical Society of America. 2021;149(4):2559-2572, ISSN: 0001-4966. DOI: 10.1121/10.0004128

[49] Katz BF, Parseihian G. Perceptually based head-related transfer function database optimization. The Journal of the Acoustical Society of America. 2012; 131(2):EL99-EL105. DOI: 10.1121/1.3672641

[50] Baumgartner R, Majdak P, Laback B. Modeling sound-source localization in sagittal planes for human listeners. The Journal of the Acoustical Society of America. 2014;136(2): 791-802. DOI: 10.1121/1.4887447

[51] Xie B, Zhong X, He N. Typical data and cluster analysis on head-related transfer functions from chinese subjects. Applied Acoustics. 2015;94:1-13. DOI: 10.1016/j.apacoust.2015.01.022

[52] Toppila E, Pyykkö I, Starck J. Age and noise-induced hearing loss. Scandinavian Audiology. 2001;30(4): 236-244. DOI: 10.1080/01050390152704751

[53] Klumpp RG, Eady HR. Some measurements of interaural time difference thresholds. The Journal of the Acoustical Society of America. 1956;28: 859-860. DOI: 10.1121/1.1908493

[54] Blauert J. Spatial hearing. In: The Psychophysics of Human Sound Localization. Cambridge, MA: The MIT Press; 1997

[55] Raykar VC, Duraiswami R, Yegnanarayana B. Extracting the frequencies of the pinna spectral

notches in measured head related impulse responses. The Journal of the Acoustical Society of America. 2005; **118**(1):364-374. DOI: 10.1121/1.1923368

[56] Takemoto H, Mokhtari P, Kato H, Nishimura R, Iida K. Mechanism for generating peaks and notches of head-related transfer functions in the median plane. The Journal of the Acoustical Society of America. 2012;**132**(6): 3832-3841. DOI: 10.1121/1.4765083

[57] Algazi VR, Duda RO, Duraiswami R, Gumerov NA, Tang Z. Approximating the head-related transfer function using simple geometric models of the head and torso. The Journal of the Acoustical Society of America. 2002;**112**(5): 2053-2064. DOI: 10.1121/1.1508780

[58] Macpherson EA, Middlebrooks JC. Vertical-plane sound localization probed with ripple-spectrum noise. The Journal of the Acoustical Society of America. 2003;**114**(1):430-445. DOI: 10.1121/ 1.1582174

[59] Goupell MJ, Majdak P, Laback B. Median-plane sound localization as a function of the number of spectral channels using a channel vocoder. The Journal of the Acoustical Society of America. 2010;**127**(2):990-1001. DOI: 10.1121/1.3283014

[60] Kulkarni A, Colburn HS. Role of spectral detail in sound-source localization. Nature. 1998;**396**(6713): 747-749. DOI: 10.1038/25526

[61] Senova MA, McAnally KI, Martin RL. Localization of virtual sound as a function of head-related impulse response duration. Journal of the Audio Engineering Society. 2002;**50**(1/ 2):57-66

[62] Thavam S, Dietz M. Smallest perceivable interaural time differences. The Journal of the Acoustical Society of America. 2019;**145**(1):458-468. DOI: 10.1121/1.5087566

[63] Andreopoulou A, Katz BF. Identification of perceptually relevant methods of inter-aural time difference estimation. The Journal of the Acoustical Society of America. 2017; **142**(2):588-598. DOI: 10.1121/1.4996457

[64] Katz BF, Noisternig M. A comparative study of interaural time delay estimation methods. The Journal of the Acoustical Society of America. 2014;**135**(6):3530-3540. DOI: 10.1121/ 1.4875714

[65] Algazi R, Avendano C, Duda RO. Estimation of a spherical-head model from anthropometry. Journal of the Audio Engineering Society. 2001;**49**: 472-479

[66] Zhang W, Abhayapala TD, Kennedy RA, Duraiswami R. Insights into head-related transfer function: Spatial dimensionality and continuous representation. The Journal of the Acoustical Society of America. 2010; **127**(4):2347-2357. DOI: 10.1121/1.3336399

[67] Bomhardt R, de la Fuente Klein M, Fels J. A high-resolution head-related transfer function and three-dimensional ear model database. In: Proceedings of Meetings on Acoustics 172ASA. Vol. 29. Illinois, United States: ASA; 2016. p. 050002. DOI: 10.1121/2.0000467

[68] Carpentier T, Bahu H, Noisternig M, Warusfel O. Measurement of a head-related transfer function database with high spatial resolution. In: 7th Forum Acusticum (EAA). Ukraine: EAA; 2014

[69] Jin CT, Guillon P, Epain N, Zolfaghari R, Van Schaik A, Tew AI, et al. Creating the Sydney York morphological and acoustic recordings of ears database. IEEE Transactions on Multimedia. 2013;**16**(1):37-46. DOI: 10.1109/TMM.2013.2282134

[70] Mills AW. On the minimum audible angle. The Journal of the Acoustical

Society of America. 1958;**30**(4):237-246. DOI: 10.1121/1.1909553

[71] Wersényi G. HRTFs in human localization: Measurement, spectral evaluation and practical use in virtual audio environment. Dissertation. Cottbus, Germany: Brandenburg University of Technology; 2002

[72] Zhong X, Xie B, et al. Head-related transfer functions and virtual auditory display. In: Soundscape Semiotics-Localization and Categorization. Plantation, FL, United States: J. Ross Publishing; 2014. p. 1. DOI: 10.5772/56907

[73] Makous JC, Middlebrooks JC. Two-dimensional sound localization by human listeners. The Journal of the Acoustical Society of America. 1990; **87**(5):2188-2200. DOI: 10.1121/1.399186

[74] Middlebrooks JC. Spectral shape cues for sound localization. In: Binaural and Spatial Hearing in Real and Virtual Environments. New York: Psychology Press; 1997. pp. 77-97

[75] Middlebrooks JC. Virtual localization improved by scaling nonindividualized external-ear transfer functions in frequency. The Journal of the Acoustical Society of America. 1999; **106**(3):1493-1510. DOI: 10.1121/1.427147

[76] Perrott DR, Saberi K. Minimum audible angle thresholds for sources varying in both elevation and azimuth. The Journal of the Acoustical Society of America. 1990;**87**(4):1728-1731, ISSN: 0001-4966. DOI: 10.1121/1.399421

[77] Middlebrooks JC, Green DM. Sound localization by human listeners. Annual Review of Psychology. 1991;**42**(1): 135-159. DOI: 10.1146/annurev. ps.42.020191.001031

[78] Poirier P, Miljours S, Lassonde M, Lepore F. Sound localization in acallosal human listeners. Brain. 1993;**116**(1): 53-69. DOI: 10.1093/brain/116.1.53

[79] Voss P, Lassonde M, Gougoux F, Fortin M, Guillemot J-P, Lepore F. Early- and late-onset blind individuals show supra-normal auditory abilities in far-space. Current Biology. 2004; **14**(19):1734-1738. DOI: 10.1016/j. cub.2004.09.051

[80] Senn P, Kompis M, Vischer M, Haeusler R. Minimum audible angle, just noticeable interaural differences and speech intelligibility with bilateral cochlear implants using clinical speech processors. Audiology and Neurotology. 2005;**10**(6):342-352. DOI: 10.1159/000087351

[81] Pulkki V. Localization of amplitude-panned virtual sources II: Two- and three-dimensional panning. Journal of the Audio Engineering Society. 2001; **49**(4):753-767

[82] Bremen P, van Wanrooij MM, van Opstal AJ. Pinna cues determine orienting response modes to synchronous sounds in elevation. Journal of Neuroscience. 2010;**30**(1): 194-204. DOI: 10.1523/JNEUROSCI.2982-09.2010

[83] Brimijoin WO, Akeroyd MA. The moving minimum audible angle is smaller during self motion than during source motion. Frontiers in Neuroscience. 2014;**8**:273. DOI: 10.3389/fnins.2014.00273

[84] Begault DR, Wenzel EM, Anderson MR. Direct comparison of the impact of head tracking, reverberation, and individualized head-related transfer functions on the spatial perception of a virtual speech source. Journal of the Audio Engineering Society. 2001; **49**(10):904-916

[85] Stitt P, Hendrickx E, Messonnier J, Katz B. The role of head tracking in binaural rendering. In: 29th

Tonmeistertagung, International VDT Convention. Germany: CCN Cologne; 2016

[86] Urbanietz C, Enzner G. Binaural rendering of dynamic head and sound source orientation using high-resolution HRTF and retarded time. In: 2018 IEEE International Conference on Acoustics, Speech and Signal Processing (ICASSP). Calgary, AB, Canada: IEEE; 2018. pp. 566-570. DOI: 10.1109/ICASSP.2018.8461343

[87] Pörschmann C, Arend JM. Obtaining dense HRTF sets from sparse measurements in reverberant environments. In: Audio Engineering Society Conference: 2019 AES International Conference on Immersive and Interactive Audio. New York, New York, United States: Audio Engineering Society; 2019

[88] Algazi VR, Duda RO, Thompson DM, Avendano C. The CIPIC HRTF database. In: Proceedings of the 2001 IEEE Workshop on the Applications of Signal Processing to Audio and Acoustics (Cat. No.01TH8575). New York: IEEE; 2001. pp. 99-102. DOI: 10.1109/ASPAA.2001.9695520

[89] Pelzer R, Dinakaran M, Brinkmann F, Lepa S, Grosche P, Weinzierl S. Head-related transfer function recommendation based on perceptual similarities and anthropometric features. The Journal of the Acoustical Society of America. 2020; **148**(6):3809-3817. DOI: 10.1121/10.0002884

[90] Ziegelwanger H, Reichinger A, Majdak P. Calculation of listener-specific head-related transfer functions: Effect of mesh quality. In: Proceedings of Meetings on Acoustics. Vol. 19. Montreal, Canada. 2013. p. 050017. DOI: 10.1121/1.4799868

[91] Gardner MB, Gardner RS. Problem of localization in the median plane:

Effect of pinnae cavity occlusion. The Journal of the Acoustical Society of America. 1973;**53**(2):400-408. DOI: 10.1121/1.1913336

[92] Nelson PA, Kahana Y. Spherical harmonics, singular-value decomposition and head-related transfer function. Journal of Sound and Vibration. 2001;**239**:607-637. DOI: 10.1006/jsvi.2000.3227

[93] Shaw EAG. The external ear. In: Keidel WD, Neff WD, editors. Auditory System. Vol. 5/1. Berlin Heidelberg, ISBN: 978-3-642-65831-0 978-3-642-65829-7: Springer; 1974. pp. 455-490. DOI: 10.1007/978-3-642-65829-7_14

[94] Brinkmann F. The FABIAN head-related transfer function data base. Berlin: Technische Universität Berlin; 2017. DOI: 10.14279/depositonce-5718

[95] Brinkmann F, Dinakaran M, Pelzer R, Grosche P, Voss D, Weinzierl S. A cross-evaluated database of measured and simulated HRTFs including 3D head meshes, anthropometric features, and headphone impulse responses. Journal of the Audio Engineering Society. 2019; **67**(9):705-718. DOI: 10.17743/jaes.2019.0024

[96] Ghorbal S, Bonjour X, Séguier R. Computed HRIRs and ears database for acoustic research. In: Audio Engineering Society Convention 148. New York, New York, United States: Audio Engineering Society; 2020

[97] Katz BF. Acoustic absorption measurement of human hair and skin within the audible frequency range. The Journal of the Acoustical Society of America. 2000;**108**(5 Pt 1):2238-2242. DOI: 10.1121/1.1314319

[98] Treeby BE, Pan J, Paurobally RM. An experimental study of the acoustic impedance characteristics of human

hair. The Journal of the Acoustical Society of America. 2007;**122**(4): 2107-2117. DOI: 10.1121/1.2773946

[99] Brinkmann F, Lindau A, Weinzierl S. On the authenticity of individual dynamic binaural synthesis. The Journal of the Acoustical Society of America. 2017;**142**(4):1784-1795, ISSN: 0001-4966. DOI: 10.1121/1.5005606

[100] Brinkmann F, Lindau A, Weinzierl S, Müller-Trapet M, Opdam R, Vorländer M, et al. A high resolution and full-spherical head-related transfer function database for different head-above-torso orientations. Journal of the Audio Engineering Society. 2017;**65**(10):841-848. DOI: 10.17743/jaes.2017.0033

[101] Ziegelwanger H, Majdak P, Kreuzer W. Numerical calculation of listener-specific head-related transfer functions and sound localization: Microphone model and mesh discretization. The Journal of the Acoustical Society of America. 2015; **138**(1):208-222, ISSN: 0001-4966. DOI: 10.1121/1.4922518

[102] Zotkin DN, Duraiswami R, Grassi E, Gumerov NA. Fast head-related transfer function measurement via reciprocity. The Journal of the Acoustical Society of America. 2006;**120**(4):2202-2215. DOI: 10.1121/1.2207578

[103] Carlile S, Leong P, Hyams S. The nature and distribution of errors in sound localization by human listeners. Hearing Research. 1997;**114**(1–2): 179-196. DOI: 10.1016/S0378-5955(97) 00161-5

[104] Masiero B, Pollow M, Fels J. Design of a fast broadband individual head-related transfer function measurement system. Vol. 97. Hirzel: Acustica; 2011. pp. 136-136

[105] Bau D, Lübeck T, Arend JM, Dziwis D, Pörschmann C. Simplifying

head-related transfer function measurements: A system for use in regular rooms based on free head movements. In: 8th International Conference of Immersive and 3D Audio. Bologna, Italy: I3DA; 2021

[106] Reijniers J, Partoens B, Steckel J, Peremans H. HRTF measurement by means of unsupervised head movements with respect to a single fixed speaker. Vol. 8. London: IEEE Access; 2020. pp. 92287-92300, ISSN: 2169–3536. DOI: 10.1109/ ACCESS.2020.2994932

[107] Fukudome K, Suetsugu T, Ueshin T, Idegami R, Takeya K. The fast measurement of head related impulse responses for all azimuthal directions using the continuous measurement method with a servo-swiveled chair. Applied Acoustics. 2007;**68**(8):864-884. DOI: 10.1016/j.apacoust.2006.09.009

[108] He J, Ranjan R, Gan W-S, Chaudhary NK, Hai ND, Gupta R. Fast continuous measurement of HRTFs with unconstrained head movements for 3d audio. Journal of the Audio Engineering Society. 2018;**66**(11): 884-900. DOI: 10.17743/jaes.2018.0050

[109] Richter J-G, Fels J. On the influence of continuous subject rotation during high-resolution head-related transfer function measurements. IEEE/ ACM Transactions on Audio, Speech, and Language Processing. 2019;**27**(4): 730-741. DOI: 10.1109/TASLP.2019. 2894329

[110] Pulkki V, Laitinen M-V, Sivonen V. HRTF measurements with a continuously moving loudspeaker and swept sines. In: Audio Engineering Society Convention 128. New York, New York, United States: Audio Engineering Society; 2010

[111] Kabzinski T, Jax P. Towards faster continuous multi-channel HRTF measurements based on learning system

models. In: 2022 IEEE International Conference on Acoustics, Speech and Signal Processing (ICASSP). Singapore: IEEE; 2021 arXiv preprint arXiv: 2110.03630

[112] Majdak P, Balazs P, Laback B. Multiple exponential sweep method for fast measurement of head-related transfer functions. Journal of the Audio Engineering Society. 2007;55:623-637

[113] Middlebrooks JC, Makous JC, Green DM. Directional sensitivity of sound-pressure levels in the human ear canal. The Journal of the Acoustical Society of America. 1989;86(1):89-108. DOI: 10.1121/1.398224

[114] Wightman F, Kistler D, Foster S, Abel J. A comparison of head-related transfer functions measured deep in the ear canal and at the ear canal entrance. In: 17th Midwinter Meeting of the Association for Research in Otolaryngology. Vol. 71. Montreal: ARO; 1995

[115] Zahorik P. Limitations in using golay codes for head-related transfer function measurement. The Journal of the Acoustical Society of America. 2000; 107(3):1793-1796. DOI: 10.1121/1.428579

[116] Dietrich P, Masiero B, Vorländer M. On the optimization of the multiple exponential sweep method. Journal of the Audio Engineering Society. 2013;61(3):113-124

[117] Armstrong C, Thresh L, Murphy D, Kearney G. A perceptual evaluation of individual and non-individual HRTFs: A case study of the SADIE II database. Applied Sciences. 2018;8(11):2029. DOI: 10.3390/app8112029

[118] Denk F, Kollmeier B, Ewert SD. Removing reflections in semianechoic impulse responses by frequency-dependent truncation. Journal of the Audio Engineering Society. 2018;66(3): 146-153. DOI: 10.17743/jaes.2018.0002

[119] Kistler DJ, Wightman FL. A model of head-related transfer functions based on principal components analysis and minimum-phase reconstruction. The Journal of the Acoustical Society of America. 1992;91(3):1637-1647. DOI: 10.1121/1.402444

[120] Kohlrausch A, Breebaart J. Perceptual (ir) relevance of HRTF magnitude and phase spectra. In: Audio Engineering Society Convention 110. New York, New York, United States: Audio Engineering Society; 2001

[121] Bergman DR. Computational Acoustics: Theory and Implementation. Hoboken, New Jersey, United States: John Wiley & Sons; 2018

[122] Marburg S. Six boundary elements per wavelength. Is that enough? Journal of Computational Acoustics. 2002;10: 25-51. DOI: 10.1142/S0218396X0 2001401

[123] Botsch M, Kobbelt L. A remeshing approach to multiresolution modeling. In: Proceedings of the 2004 Eurographics/ACM SIGGRAPH Symposium on Geometry Processing. New York, NY, United States: Association for Computing Machinery; 2004. pp. 185-192. DOI: 10.1145/ 1057432.1057457

[124] Reichinger A, Majdak P, Sablatnig R, Maierhofer S. Evaluation of methods for optical 3-D scanning of human pinnas. In: Proceedings of the 3D Vision Conference. Seattle, WA: IEEE; 2013. pp. 390-397. DOI: 10.1109/ 3DV.2013.58

[125] Dinakaran M, Brinkmann F, Harder S, Pelzer R, Grosche P, Paulsen RR, et al. Perceptually motivated analysis of numerically simulated head-related transfer functions generated by various 3d surface scanning systems. In: 2018 IEEE International Conference on Acoustics, Speech and Signal Processing (ICASSP).

Calgary, Alberta, Canada: IEEE; 2018. pp. 551-555. DOI: 10.1109/ICASSP.2018.8461789

[126] Greff R, Katz BF. Round robin comparison of HRTF simulation systems: Preliminary results. In: Audio Engineering Society Convention 123, Page Convention Paper 7188. New York, New York, United States: Audio Engineering Society; 2007

[127] Dellepiane M, Pietroni N, Tsingos N, Asselot M, Scopigno R. Reconstructing head models from photographs for individualized 3d-audio processing. In: Computer Graphics Forum. Vol. 27. Hoboken, New Jersey, United States: Wiley Online Library; 2008. pp. 1719-1727. DOI: 10.1111/j.1467-8659.2008.01316.x

[128] Iida K, Nishiyama O, Aizaki T. Estimation of the category of notch frequency bins of the individual head-related transfer functions using the anthropometry of the listener's pinnae. Applied Acoustics. 2021;**177**:107929. DOI: 10.1016/j.apacoust.2021.107929

[129] Pollack K, Brinkmann F, Majdak P, Kreuzer W. Von Fotos zu personalisierter räumlicher Audiowiedergabe [from photos to personalised spatial audio playback]. e & i Elektrotechnik und Informationstechnik. 2021;**138**(3):1-6. DOI: 10.1007/s00502-021-00891-4

[130] Ullman S, Brenner S. The interpretation of structure from motion. Proceedings of the Royal Society of London. Series B. Biological Sciences. 1979;**203**(1153):405-426. DOI: 10.1098/rspb.1979.0006 Publisher: Royal Society

[131] Sommerfeld A. Partial Differential Equations in Physics. Cambridge, Massachusetts, United States: Academic Press; 1949

[132] Turner MJ, Clough RW, Martin HC, Topp L. Stiffness and deflection analysis of complex

structures. Journal of the Aeronautical Sciences. 1956;**23**(9):805-823. DOI: 10.2514/8.3664

[133] Bériot H, Prinn A, Gabard G. Efficient implementation of high-order finite elements for Helmholtz problems. International Journal for Numerical Methods in Engineering. 2016;**106**(3): 213-240. DOI: 10.1002/nme.5172

[134] Gabard G, Bériot H, Prinn A, Kucukcoskun K. Adaptive, high-order finite-element method for convected acoustics. AIAA Journal. 2018;**56**(8): 3179-3191. DOI: 10.2514/1.J057054

[135] Ueberhuber CW. Numerical Computation 1: Methods, Software, and Analysis. Vol. 16. Berlin, Germany: Springer Science & Business Media; 1997

[136] Beriot H, Modave A. An automatic perfectly matched layer for acoustic finite element simulations in convex domains of general shape. International Journal for Numerical Methods in Engineering. 2021;**122**(5):1239-1261. DOI: 10.1002/nme.6560

[137] Farahikia M, Su QT. Optimized finite element method for acoustic scattering analysis with application to head-related transfer function estimation. Journal of Vibration and Acoustics. 2017;**139**(3):034501. DOI: 10.1115/1.4035813

[138] Harder S, Paulsen RR, Larsen M, Laugesen S, Mihocic M, Majdak P. A framework for geometry acquisition, 3-D printing, simulation, and measurement of head-related transfer functions with a focus on hearing-assistive devices. Computer Aided Design. 2016;**75-76**:39-46, ISSN: 0010-4485. DOI: 10.1016/j.cad.2016.02.006

[139] Huttunen T, Seppälä ET, Kirkeby O, Kärkkäinen A, Kärkkäinen L. Simulation of the transfer function for a head-and-torso model

over the entire audible frequency range. Journal of Computational Acoustics. 2007;**15**(04):429-448. DOI: 10.1142/ S0218396X07003469

[140] Kahana Y. Numerical Modelling of the Head-Related Transfer Function. Southampton, UK: University of Southampton; 2000

[141] Ma F, Wu JH, Huang M, Zhang W, Hou W, Bai C. Finite element determination of the head-related transfer function. Journal of Mechanics in Medicine and Biology. 2015;**15**(05): 1550066. DOI: 10.1142/ S0219519415500669

[142] Yee K. Numerical solution of initial boundary value problems involving Maxwell's equations in isotropic media. IEEE Transactions on Antennas and Propagation. 1966;**14**(3):302-307. DOI: 10.1109/TAP.1966.1138693

[143] Botts J, Savioja L. Spectral and pseudospectral properties of finite difference models used in audio and room acoustics. IEEE/ACM Transactions on Audio, Speech, and Language Processing. 2014;**22**(9): 1403-1412. DOI: 10.1109/ TASLP.2014.2332045

[144] Häggblad J, Runborg O. Accuracy of staircase approximations in finite-difference methods for wave propagation. Numerische Mathematik. 2014;**128**(4):741-771. DOI: 10.1007/ s00211-014-0625-1

[145] Prepeliţă ST, Geronazzo M, Avanzini F, Savioja L. Influence of voxelization on finite difference time domain simulations of head-related transfer functions. The Journal of the Acoustical Society of America. 2016; **139**(5):2489-2504. DOI: 10.1121/ 1.4947546

[146] Prepeliţă ST, Gómez Bolaños J, Geronazzo M, Mehra R, Savioja L. Pinna-related transfer functions and

lossless wave equation using finite-difference methods: Verification and asymptotic solution. The Journal of the Acoustical Society of America. 2019; **146**(5):3629-3645. DOI: 10.1121/ 1.5131245

[147] Prepeliţă ST, Gómez Bolaños J, Geronazzo M, Mehra R, Savioja L. Pinna-related transfer functions and lossless wave equation using finite-difference methods: Validation with measurements. The Journal of the Acoustical Society of America. 2020; **147**(5):3631-3645. DOI: 10.1121/ 10.0001230

[148] Botteldooren D. Acoustical finite-difference time-domain simulation in a quasi-cartesian grid. The Journal of the Acoustical Society of America. 1994; **95**(5):2313-2319. DOI: 10.1121/1.409866

[149] Willemsen S, Bilbao S, Ducceschi M, Serafin S. Dynamic grids for finite-difference schemes in musical instrument simulations. In: 24th International Conference on Digital Audio Effects. Vienna, Austria: DAFX; 2021. pp. 144-151

[150] Bilbao S. Modeling of complex geometries and boundary conditions in finite difference/finite volume time domain room acoustics simulation. IEEE Transactions on Audio, Speech, and Language Processing. 2013;**21**(7): 1524-1533. DOI: 10.1109/ TASL.2013.2256897

[151] Bilbao S, Hamilton B. Passive volumetric time domain simulation for room acoustics applications. The Journal of the Acoustical Society of America. 2019;**145**(4):2613-2624. DOI: 10.1121/ 1.5095876

[152] Bilbao S, Hamilton B, Botts J, Savioja L. Finite volume time domain room acoustics simulation under general impedance boundary conditions. IEEE/ ACM Transactions on Audio, Speech, and Language Processing. 2015;**24**(1):

161-173. DOI: 10.1109/
TASLP.2015.25000180

[153] Peiró, J. Sherwin S. Finite
difference, finite element and finite
volume methods for partial differential
equations. In Handbook of Materials
Modeling. Berlin, Germany: Springer;
2005. pp. 2415–2446. DOI: 10.1007/
978-1-4020-3286-8_127

[154] Mokhtari P, Takemoto H,
Nishimura R, Kato H. Frequency and
amplitude estimation of the first peak of
head-related transfer functions from
individual pinna anthropometry. The
Journal of the Acoustical Society of
America. 2015;**137**(2):690-701. DOI:
10.1121/1.4906160

[155] Xiao T, Huo Liu Q. Finite
difference computation of head-related
transfer function for human hearing.
The Journal of the Acoustical Society of
America. 2003;**113**(5):2434-2441, ISSN:
0001-4966. DOI: 10.1121/1.1561495

[156] Gumerov NA, O'Donovan AE,
Duraiswami R, Zotkin DN.
Computation of the head-related
transfer function via the fast multipole
accelerated boundary element method
and its spherical harmonic
representation. The Journal of the
Acoustical Society of America. 2010;
127(1):370-386. DOI: 10.1121/1.3257598

[157] Galerkin BG. Rods and plates.
Series occurring in various questions
concerning the elastic equilibrium of
rods and plates. Engineers Bulletin
(Vestnik Inzhenerov). 1915;**19**:897-908

[158] Nyström EJ. Über die praktische
Auflösung von Integralgleichungen mit
Anwendungen auf Randwertaufgaben
[about the practical solution of integral
equations with applications to boundary
value problems]. Acta Mathematica. 1930;
54:185-204. DOI: 10.1007/BF02547521

[159] Sauter S, Schwab S. Boundary
Element Methods. Berlin, Germany:
Springer; 2011

[160] Arnold DN, Wendland WL.
Collocation versus Galerkin procedures
for boundary integral methods. In:
Brebbia CA, editor. Boundary Element
Methods in Engineering. Berlin,
Germany ISBN: 978-3-662-11275-5:
Springer International Publishing; 1982.
DOI: 10.1007/978-3-662-11273-1_2

[161] Duffy MG. Quadrature over a
pyramid or cube of integrands with a
singularity at a vertex. SIAM Journal on
Numerical Analysis. 1982;**19**(6):
1260-1262. DOI: 10.1137/0719090

[162] Krishnasamy G, Schmerr L,
Rudolphi T, Rizzo F. Hypersingular
boundary integral equations: Some
applications in acoustic and elastic wave
scattering. Transactions of the ASME.
1990;**57**:404-414. DOI: 10.1115/
1.2892004

[163] Coifman R, Rokhlin V,
Wandzura S. The fast multipole method
for the wave equations: A pedestrian
prescription. IEEE Antennas and
Propagation Magazine. 1993;**35**(3):7-12,
ISSN: 1045-9243. DOI: 10.1109/
74.250128

[164] Hackbusch W. Hierarchical
Matrices: Algorithms and Analysis.
Berlin, Heidelberg: Springer; 2015. DOI:
10.1007/978-3-662-47324-5

[165] Kreuzer W, Majdak P, Chen Z. Fast
multipole boundary element method to
calculate head-related transfer functions
for a wide frequency range. The Journal
of the Acoustical Society of America.
2009;**126**(3):1280-1290. DOI: 10.1121/
1.3177264

[166] Saad Y. Iterative Methods for
Sparse Linear Systems. New Delhi,
India: SIAM; 2003

[167] Burton AJ, Miller GF. The
application of integral equation methods
to the numerical solution of some
exterior boundary-value problems.
Proceedings of the Royal Society of

London A. Mathematical and Physical Sciences. 1971;**323**(1553):201-210, ISSN: 0080-4630. DOI: 10.1098/rspa.1971.0097

[168] Katz BF. Boundary element method calculation of individual head-related transfer function. I. Rigid model calculation. The Journal of the Acoustical Society of America. 2001;**110**(5 Pt 1):2440-2448. DOI: 10.1121/1.1412440

[169] Katz BF. Boundary element method calculation of individual head-related transfer function. II. Impedance effects and comparisons to real measurements. The Journal of the Acoustical Society of America. 2001;**110**(5 Pt 1):2449-2455. DOI: 10.1121/1.1412441

[170] Otani M, Ise S. A fast calculation method of the head-related transfer functions for multiple source points based on the boundary element method. Acoustical Science and Technology. 2003;**24**(5):259-266. DOI: 10.1250/ast.24.259

[171] Otani M, Ise S. Fast calculation system specialized for head-related transfer function based on boundary element method. The Journal of the Acoustical Society of America. 2006;**119**(5 Pt 1):2589-2598, ISSN: 0001-4966. DOI: 10.1121/1.2191608

[172] Ziegelwanger H, Kreuzer W, Majdak P. Mesh2HRTF: Open-source software package for the numerical calculation of head-related transfer functions. In Proceedings of the 22nd International Congress on Sound and Vibration, 1–8, IEEE Florence, IT. 2015. DOI: 10.13140/RG.2.1.1707.1128

[173] Fink KJ, Ray L. Individualization of head related transfer functions using principal component analysis. Applied Acoustics. 2015;**87**:162-173. DOI: 10.1016/j.apacoust.2014.07.005

[174] Xie B, Zhong X, Rao D, Liang Z. Head-related transfer function database and its analyses. Science in China Series G: Physics, Mechanics and Astronomy. 2007;**50**(3):267-280, ISSN: 1672-1799, 1862-2844. DOI: 10.1007/s11433-007-0018-x

[175] Nishino T, Inoue N, Takeda K, Itakura F. Estimation of HRTFs on the horizontal plane using physical features. Applied Acoustics. 2007;**68**(8):897-908, ISSN: 0003-682X. DOI: 10/dr4tg3

[176] Xie B. Head-Related Transfer Function and Virtual Auditory Display. Plantation, FL, United States: J. Ross Publishing; 2013

[177] Gromov M. Metric structures for Riemannian and non-Riemannian spaces. Bulletin of the American Mathematical Society. 2001;**38**:353-363

[178] Hebrank J, Wright D. Are two ears necessary for localization of sound sources on the median plane? The Journal of the Acoustical Society of America. 1974;**56**(3):935-938. DOI: 10.1121/1.1903351

Section 2

Perception

HRTF Performance Evaluation: Methodology and Metrics for Localisation Accuracy and Learning Assessment

David Poirier-Quinot, Martin S. Lawless, Peter Stitt and Brian F.G. Katz

Abstract

Through a review of the current literature, this chapter defines a methodology for the analysis of HRTF localisation performance, as applied to assess the quality of an HRTF selection or learning program. A case study is subsequently proposed, applying this methodology to a cross-comparison on the results of five contemporary experiments on HRTF learning. The objective is to propose a set of steps and metrics to allow for a systematic assessment of participant performance (baseline, learning rates, foreseeable performance plateau limits, etc.) to ease future inter-study comparisons.

Keywords: spatial hearing, binaural, localisation accuracy, evaluation, HRTF selection, HRTF training

1. Introduction

If you reached this point, you are probably familiar with the concept of binaural rendering. You likely also know that it is used for producing spatial sound over headphones in most of today's personal mixed reality experiences. While conceptually sound, binaural rendering is subject to several limitations in practice, some of them leading users to perceive distorted versions of the encoded 3D scene. Those distortions range from slight localisation blur to critical scenarios where auditory events are perceived on the opposite hemisphere from their actual position. Researchers have been working on techniques to address this problem of binaural localisation accuracy for some time now. To establish the benefit of these techniques, they predominantly, and quite naturally, rely on localisation performance evaluations.

The problem that concerns us here is that there is no standard for said evaluation. As a consequence, fully appreciating the value of a technique often requires careful reading and interpretation of both protocols and associated results. This becomes truly problematic when comparing the results of several studies, where differences in protocol and evaluation metrics make for complicated analysis at best, simply impossible in some cases. Without inter-study comparison, it becomes hard to reach any conclusion on the overall and added value of an HRTF selection, synthesis, or learning method. The objective of this chapter is to lay the foundations of such a standard.

1.1 Context

One of the most frequent causes of auditory space distortion in binaural rendering is related to the use of *non-individual* Head Related Transfer Functions (HRTF)[1]. An HRTF is a collection of filter pairs that, applied to a mono signal, modify it so that it has the same characteristics as if it had physically been travelling from a specific point in space to our ears. The term HRTF refers to the set of filter pairs, each corresponding to a different source position, typically forming a sphere of fixed radius around the listener. When sound travels to our ears, the acoustic wave interactions with our morphology causes deformations in the perceived signal. From childhood, our brain learned to interpret these acoustic cues as different source positions. Since there exist many variations of ear, head, and torso shapes that each deform the sound differently, so too are there variations in HRTFs. While we are quite adept at sound localisation with our own ears and our own HRTF, the problem arises when we start using someone else's.

In practice, most users will end up experiencing binaural rendering using an HRTF that is not their own, as in the case of a non-individual HRTF, generally taken from an existing database. Presently, measuring an individual's HRTF most often requires specific equipment and access to an anechoic room. Methods exist to simulate an HRTF from geometrical head scans or morphological data, but they suffer the same drawbacks: the techniques are either too costly or burdensome to implement in practical scenarios, or they produce HRTFs that do not exactly match the individual users. As mentioned, using a non-individual HRTF, which the brain has not trained with, often results in distortions of the perceived auditory space. Researchers have been working on this issue, proposing new simulation methods, HRTF selection processes, and even HRTF training programs focused on the reduction of these distortions.

Naturally, all these lines of research end up using a localisation evaluation task to assess the benefit of new techniques. As mentioned above, there exists no standard method for this evaluation, hindering results appraisal and inter-study comparisons.

1.2 Chapter scope and organisation

The objective of this chapter is to outline a set of metrics and propose a methodology to assess localisation performance in the context of HRTF selection and training programs. While the tools proposed can be applied to other contexts, they were designed with HRTF training in mind as not only do they assess instantaneous performance but also performance *evolution*, adding another dimension to the analysis workflow.

Section 2 presents a state of the art of evaluation metrics used to assess localisation accuracy in previous studies. Section 3 introduces the proposed methodology and the set of metrics on which it is built. Section 4 is a case-study, using

---

[1] We use the term *individual* to identify the HRTF of the user, *individualised* or *personalised* to indicated an HRTF modified or selected to best accommodate the user, and *non-individual* or *non-individualised* to indicate an HRTF that has not been tailored to the user. A so-called *generic* or *dummy-head* HRTF are specific instances of non-individual HRTFs.

the methodology to re-analyse and compare the results of five contemporary experiments on HRTF learning. Section 4 concludes this chapter.

2. State of the art

This section presents and discusses a variety of metrics and methods of analysis introduced in previous studies for the evaluation of auditory localisation performance, in the context of HRTF selection and learning. Further, it discusses what aspect of the data or human behaviour is highlighted by each metric.

2.1 Analysis based on angular distances

The majority of the metrics used in the literature to assess localisation performance are derived from the angular distance from the source position to the participant's response. This section discusses the most common of these metrics, their interpretation, and limitations. It builds upon the work presented in Letowski and Letowski [1].

2.1.1 Egocentric coordinate systems

Many auditory localisation tasks have participants indicating perceived target locations *around them*. As such, egocentric coordinate systems are a logical choice for the assessment of pointing errors. The *spherical* coordinate system, illustrated in **Figure 1a**, uses axes of azimuth and elevation angles. As most researchers are familiar with this coordinate system, it provides an intuitive framework to view and present results.

Alternatively, the *interaural* coordinate system has been proposed to evaluate localisation results as a more natural representation of how sound is perceived. The lateral angle, referred to as the "binaural disparity cue" by Morimoto and Aokata [2], defines *cones-of-confusion* along which the binaural cues of Interaural Level Difference (ILD) and Interaural Time Difference (ITD) are approximately constant. A cone-of-confusion is a set of positions presenting binaural cue/localisation ambiguities, that listeners may not be able to differentiate unless provided with further spectral cues or head movement information [3]. While not truly 'cones',

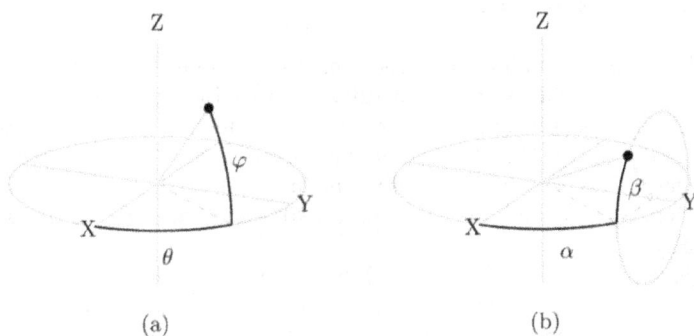

Figure 1.
(a) Spherical, and (b) interaural coordinate systems used in the methodology, for a source positioned at angles (55°, 46°) as defined in each coordinate system. Spherical azimuth angle θ is defined in [−180°:180°], elevation angle φ in [−90°:90°]. Interaural lateral angle α is defined in [−90°:90°], polar angle β in [−180°:180°]. The lateral angle used here is shifted by 90° compared to that originally defined by Morimoto and Aokata [2]. In both systems, listeners are facing X with their left ear pointing towards Y.

these constant ILD or ITD surfaces generally define a circle when the radius is fixed (see [4] for more discussion on the variation with radius of these constant-value surfaces). To maintain accepted terminology in the field, each of these circles is termed a "cone-of-confusion". The polar angle, or "spectral cue", is primarily linked with the monaural spectral cues in the HRTF. This independence of binaural and monaural cues makes the interaural coordinate system a compelling choice when assessing localisation performance, particularly when monaural cues are of special interest as in HRTF selection and learning tasks.

Other conventions have been proposed, such as the *double-pole* [5] or *three-pole* [6] coordinate systems. These systems have been designed to circumvent compression issues impacting single-pole (spherical and interaural) coordinate systems, further discussed in Section 2.1.3. They can prove very helpful for some types of data presentation [5], yet can confuse the analysis as more than one coordinate vector can be assigned to any given point in space.

2.1.2 Azimuth, elevation, lateral, and polar errors

Regardless of the coordinate system used, angular errors can be calculated using either the *signed* or *absolute difference* between target and response coordinates. The signed error will give an indication on the "localisation bias" [5] where the absolute error, more often used in the literature [7–10], provides a measure of how close a response is to the target, regardless of error direction. Computing summary statistics from these values can be a first and straightforward step to characterise both the central tendency and dispersion, or "localisation blur" [11], of participant responses [1].

Care must be taken in calculating signed and absolute errors because of the discontinuities in the azimuth and polar angles of the spherical and interaural coordinate systems. If a source is close to the discontinuity and the response crosses it (*e.g.* 179° to −179°), the calculated error will be artificially large. Likewise, summary statistics such as mean or standard deviation should also be computed away from those discontinuities. Another problem that results from working with egocentric systems is that data distributions will be warped by the sphere curvature, requiring in theory to use circular statistics when comparing statistical distributions. As discussed in [1], linear statistics can however be used in practice if the directional judgements are relatively well concentrated around a central direction.

2.1.3 Compensating for spatial compression

Both the spherical and interaural coordinate systems introduce spatial compression at their poles. In the interaural coordinate system for example, the circumference of the cone-of-confusion at 80° lateral angle is much smaller than that of a cone at 0° lateral angle. Therefore, polar angle errors at the poles (near ±90° lateral angle) are more exaggerated than near the median plane. The same problem impacts azimuth errors near the poles (near ±90° elevation angle) for the spherical coordinate system.

Previous studies have sought to avoid the spatial compression problem altogether by limiting the analysis to targets away from the poles [12]. The downside of this method is that it limits the scope of the study's conclusions because a large region of space cannot be studied. Still others have proposed compensation schemes, using for example the lateral angle to weight the response contribution to the average polar error [13–15]. Carlile et al. [13] for example weighted polar response errors using the cosine of the target lateral angle, decreasing response contributions as targets moved towards the interaural axis. This method more accurately reflects the arc length between the target and response locations on the circle, keeping in mind that this weighting does not take the lateral angle of the response into account.

2.1.4 Using directional statistics to analyse sound localisation accuracy

Due to the discontinuities and spatial compression in the angular metrics of the typical coordinate systems, some work has simply examined the distance between the participant responses and the true target positions to assess the extent of localisation error. The most basic method, the *great-circle error* used in several studies [9, 15, 16], is measured as the distance along the unit sphere between the response and target locations. The great-circle error is independent of the selected coordinate system, not affected by the issues related to discontinuity in the axes or spatial compression.

Great-circle error on its own does not provide information about the direction of the response. Paired with the *angular direction*, it becomes a vector that fully describes the difference between the response and target positions [1]. Similar to *bearing* used to navigate on the globe, angular direction is the angle between the vector of the target towards the positive pole and that of the target towards the response. This vector can be used to compute the mean position of the responses, or *centroid*, and perform directional or spherical statistics. Alternatively, the centroid of the response locations may be calculated by separately summing the x, y, and z coordinates of the responses and dividing by the resultant length [17, 18], though this method may experience some undesirable results for edge cases with widely-scattered locations on the sphere.

To perform statistical analyses of the localisation accuracy, the variance in the response locations must be quantified [19, 20]. Given the two-dimensionality of the data, previous work has used Kent distributions on a sphere [17, 21] to determine ellipses that portray the variance of the data along major and minor axes of the spread of the responses. With Kent distributions, circular statistical tests may be conducted to evaluate the significance of the distance between the centroid of the responses and the target location (such as the Rayleigh z test) or the differences between mean response locations for different conditions (such as the Watson two-sample U^2 test) [22]. Alternatively, Wightman and Kistler [18] suggest the use of the "concentration parameter" κ to characterise the variance, or "dispersion", of the response locations on the sphere.

2.1.5 Further high level metrics based on angular distances

The *spherical correlation coefficient* has been used to provide an overall measure of the correlation between target and response positions [13, 17, 18]. As with standard correlation, the spherical correlation coefficient ranges from −1 to 1, where a value of 1 is obtained for two identical data sets, and a value of −1 is obtained for two sets that are reflections of one another. By construction, the spherical correlation coefficient is invariant for global rotations between the two sets.

Rather than looking at single or mean error values to assess localisation accuracy, Hofman et al. [23] and Trapeau et al. [24] studied the linear regression between targets and responses elevation angles. Termed "elevation gain", the slope of this regression provides a higher level metric that can be used to detect compression or dilation effects in participant responses. Van Wanrooij and Van Opstal [25] extended this technique, applying the regression on target versus response azimuth as well as elevation angles. To account for azimuthal dependence of the elevation gain, they also introduced the notion of "local elevation gain", averaging elevation gain values based on a sliding azimuthal window. This metric allows the assessment of how elevation compression and dilation effects impact different regions of the sphere.

2.2 Analysis based on confusions classification

2.2.1 Confusions classification

An analysis based on angular distances alone would fail to distinguish local accuracy misinterpretations from critical space confusions, where responses are often on the opposite hemisphere from target positions. These kinds of errors are very common in studies using non-individualised HRTFs [8, 10, 26, 27], though they also occur when listening with one's own ears or HRTF [5].

One of the simplest techniques is that used by Honda et al. [28], which defines a hit-miss criterion based on a threshold great-circle error value. Though intuitive, the method does not provide much information on the nature or potential origin of the confusions.

A slightly more elaborate form of confusion classification was used by Middlebrooks [12], which flags responses as confusions when they are in a different hemisphere than that of the target. To avoid reporting small local accuracy errors as confusions for targets near the hemispheres limits, only those responses with polar angle errors greater than 90° were considered when searching for confusions. The classification thus resulted in three types of "quadrant confusions": front-back, up-down, and left-right. Majdak et al. [14] further improved the definition, introducing a weighting factor to compensate for polar angle compression near the interaural axis. A comparable strategy was adopted by Carlile et al. [13], excluding from confusion checks those targets too close to the interaural axis.

A parallel classification was proposed by Martin et al. [29], determining confusion types based on cone-of-confusion angle values rather than sphere quadrants. The classification was further refined by Yamagishi and Ozawa [30], Parseihian and Katz [8] and Zagala et al. [16], adding "precision" and "combined" confusions to the already existing confusion types. This classification is discussed in more detail in Section 3.1.4.

2.2.2 Separating angular and confusions errors contributions

Given the relatively high incidence of front-back confusions in non-individual HRTF localisation tasks, results often exhibit a bi-modal distribution [10]. Analyses applied to data that contain a large portion of front-back confusions will have large variance and potentially inaccurate averages. The other confusion types also have a similar, if somewhat less characteristic, impact on the data, artificially inflating localisation errors. As such, it is common practice to split the data to analyse confusions separately from *local* performance [1, 12, 14, 31]. A potential problem with this approach is that excluding data from an analysis may result in an unbalanced data set, which limits the use of classical repeated-measures statistics.

Another approach that preserves the sample size of the data consists of 'folding' the responses into the same subspace as that of the target prior to the analysis. This technique has only ever been applied to mirror front-back confusions [18], as it may only apply to very specific circumstances and tends to inflate the power of the resulting conclusions [1].

2.3 Additional analysis methods

2.3.1 Decomposing the analysis across sphere regions

Several studies have shown variations in localisation accuracy as a function of region on the sphere due to, amongst other things, cue interpretation [3] or reporting method [32]. In these cases, decomposition schemes were used to better characterise those variations and understand their origins. As mentioned in Section 2.1.5, Van Wanrooij and Van Opstal [25] for example decomposed the analysis of elevation gain across azimuthal regions. Later, Majdak et al. [14] proposed an analysis split into hemi-fields to detect higher accuracy variations for targets in the rear region. Middlebrooks [12] applied a similar spatial decomposition to detect high variability for responses in the upper-rear quadrant, temporarily excluding them from the analysis to better assess variations in remaining regions. The principal drawback of decomposition is that it reduces the statistical power of the analysis, and can result in unbalanced data sets if responses are not evenly spread across the regions under consideration.

2.3.2 Performance evolution modelling and analysis

For the evaluation of HRTF learning, it is essential to assess the progression of participant performance over multiple sessions. On the assumption that any adaptation to an HRTF is a process with diminishing returns with repeated training sessions, localisation performances may be modelled as an exponential decay $y = y_0 \exp(-t/\tau) + c$ [15, 31]. Here y_0 is the initial performance, t is the time (training day, session, *etc.*), τ is the improvement time constant, and c is the long term performance. This model of performance over time allows for comparisons between studies, such as determining if different protocols lead to faster learning rates or if better long term performance can be achieved. If the training duration proves insufficient to reach a performance plateau/asymptote, like that seen in Stitt et al., [10], the improvement data may be better modelled using the linear form $ax + b$ [9, 31]. In addition to performance modelling, the correlation between training duration and performance metrics has been used to determine if factors other than training duration, like participant attention, should be considered to explain performance evolution [33].

Analysis of performance evolution can be performed per condition (grouping participants) [8, 10] or per participant [23]. Participant performance evaluation makes it harder to draw general conclusions, but potentially provides deeper insight into performance as not all participants exhibit the same ability to adapt to a new HRTF [24]. This adaptation capacity appears to be a function of initial HRTF affinity or "perceptual quality" [10]. For inter-study comparisons, some form of performance scaling or normalisation may first be required to compensate for such affinities, highlighting performance improvement rather than absolute value [10].

3. Methodology for assessing localisation performance

From the literature review in the previous section, a methodology is derived for assessing binaural localisation accuracy. Though it was designed with a focus on HRTF training programs, it should be applicable to any HRTF-related study interested in localisation performance assessment. Section 3.1 introduces the conventions and metrics used in the methodology, itself detailed in Section 3.2. The metrics

Name	Notion examined
Space coverage statistic	Density and homogeneity of the evaluation grid
Confusion rates	Percentage of errors resulting from cone-of-confusion or quadrant ambiguities
Great-circle error	Overall localisation accuracy
Local great-circle error	overall localisation accuracy, excluding confusions
Local lateral error	Localisation accuracy in the horizontal plane, excluding confusions
Local polar error	Localisation accuracy in the vertical plane, excluding confusions
Local azimuth error	Localisation accuracy in the horizontal plane, excluding confusions
Local elevation error	Localisation accuracy in the vertical plane, excluding confusions
Local lateral compression	Whether localisation errors are distorted systematically towards the median plane ZX, excluding confusions
Local elevation compression	Whether localisation errors are distorted systematically towards the horizontal plane XY, excluding confusions
Local lateral bias	Whether there is a systematic rotational offset on responses around the Z axis, excluding confusions
Local elevation bias	Whether there is a systematic upward offset on responses, towards positive Z, excluding confusions
Per-region metrics	Decomposition of the analysis across target regions
Local responses distribution	Whether two sets of responses, excluding confusions, belong to different spherical distributions (using Kent distribution and circular statistics)

Table 1.
Summary of the evaluation metrics used in the methodology, grouped by concept similarity.

proposed along with the notions they examine are summarised in **Table 1** at the end of this section. A MATLAB toolbox for the evaluation of all the metrics discussed here is available online[2].

3.1 Conventions and evaluation metrics

3.1.1 Coordinate systems

The methodology makes use of both spherical and interaural coordinate systems, illustrated in **Figure 1**. While the spherical coordinate system provides an intuitive perspective on the results, the interaural system has been especially designed to separate the analysis of binaural and monaural cues, as discussed in Section 2.1.1, making it a natural choice for the analysis of HRTF-related localisation performance.

3.1.2 Protocol space coverage

Space coverage is a set of metrics, sc_{angle} and sc_{shape}, designed to provide insight on the density of points tested during the localisation task, as well as on the homogeneity of their distribution on the sphere. sc_{angle} represents the density of the

---

[2] MATLAB auditory localisation evaluation toolbox: https://hal.archives-ouvertes.fr/hal-03265190.

$sc_{angle} = 18.0° \pm 0.5$ $sc_{angle} = 36.0° \pm 1.2$ $sc_{angle} = 36.0° \pm 14.5$ $sc_{angle} = 36.0° \pm 4.1$
$sc_{shape} = 0.87 \pm 0.01$ $sc_{shape} = 0.91 \pm 0.02$ $sc_{shape} = 0.80 \pm 0.09$ $sc_{shape} = 0.30 \pm 0.08$

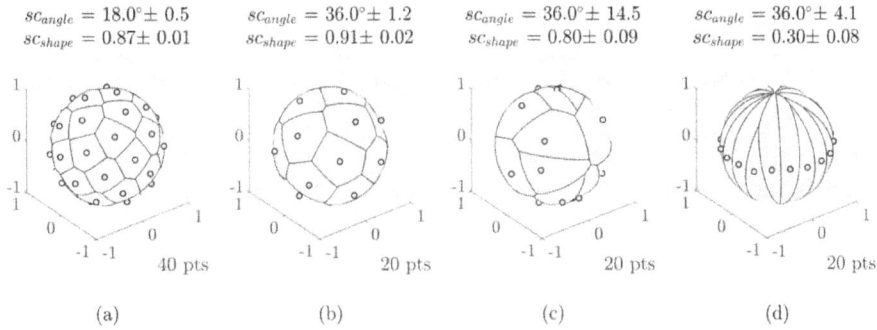

40 pts   20 pts   20 pts   20 pts

(a)   (b)   (c)   (d)

Figure 2.
Various test grids and associated space coverage statistics. (a) Homogeneous grid with large number of points, (b) homogeneous grid with small number of points, (c) non-homogeneous grid with small number of points, and (d) horizontal grid with small number of points.

evaluated positions for a given test protocol. It is is computed based on the spherical Voronoi diagram built from the evaluated positions, as the average over the solid angles of its cells [34], accompanied, $\pm$, by its standard deviation. As illustrated in **Figure 2**, denser grids result in smaller sc_{angle}, with standard deviation decreasing for increasingly homogeneous distributions.

sc_{shape} is computed as the average over the shape indices of the cells of the Voronoi diagram, defined as:

$$shape\_index = 4\pi \frac{cell\_area}{(cell\_perimeter)^2} \tag{1}$$

where the perimeter is computed as the sum of the great-circle values between the cell vertices, expressed in radians. The squared value of the perimeter, as well as a 4π normalisation factor, are used so that the final shape index value is defined in $[0, 1]$. Cells shaped as circles will have an index close to 1, whereas the index will decrease towards 0 as the cell grows into an elongated polygon. As illustrated in **Figure 2**, sc_{shape} is used in addition to sc_{angle} standard deviation to detect uneven evaluation grid distributions. Note that grid density has a negative impact on sc_{shape}: dropping from 0.91 to 0.84 for uniform grids of 20 and 80 points respectively [35].

3.1.3 Great circle error and angular direction

The great-circle error is defined as the minimum arc between the response and the true target position. This metric provides an intuitive way to assess the local localisation accuracy as the spherical distance between the responses and the target. Given xyz_{target} and $xyz_{response}$ as the vectors in Cartesian coordinates of the target and response positions respectively, the great-circle error is defined in $[0°:180°]$ as:

$$great\_circle\_error = arctan \left(\frac{\|xyz_{target} \times xyz_{response}\|}{xyz_{target} \cdot xyz_{response}} \right) \tag{2}$$

where smaller values correspond to better localisation performances.

The angular direction is coupled to the great circle to enable vector summation of target to response arcs on the sphere. The direction towards the right ear constitutes the positive pole in the interaural coordinate system. The angular direction may then be calculated from the interaural coordinates as:

$$\text{angular}_{\text{dir.}} = \arctan\left(\frac{\cos\left(\alpha_{resp}\right)\sin\left(\beta_{resp} - \beta_{target}\right)}{\cos\left(\alpha_{target}\right)\sin\left(\alpha_{resp}\right) - \sin\left(\alpha_{target}\right)\cos\left(\alpha_{resp}\right)\cos\left(\beta_{resp} - \beta_{target}\right)}\right)$$

(3)

where α is the lateral angle and β is the polar angle.

3.1.4 Confusion classification

As discussed in Section 2.2, confusion classification schemes are primarily designed to separate small localisation errors from larger errors caused by erroneous localisation behaviours typically observed in binaural localisation tasks. The scheme used in the methodology is designed around notions borrowed from both cone-of-confusion [8, 10, 16, 29] and sphere quadrant [12, 14] classifications. It separates responses into 4 categories: those near the target (*precision* errors), those opposite the target compared to the YZ plane (*front-back* errors), those within the target cone-of-confusion (*in-cone* errors), and the remainder (*off-cone* errors).

The classification is illustrated in **Figure 3a**. Responses within a 45° radius cone around the target are defined as precision errors. Responses within a 45° cone around the symmetrical of the target position regarding the YZ plane, not already classified as precision errors, are defined as front-back errors. Responses with a lateral angle within 45° of that of the target, not already classified as either precision or front-back confusions, are defined as in-cone errors. Remaining responses are defined as off-cone errors. **Figure 3b** and **c** schematically show several alternate approaches, evaluated before choosing the current method (discussed in more detail below).

The proposed 45° threshold value is somewhat arbitrary, based on a segmentation of localisation error distributions of responses from previous studies [8–10]. This value can be adapted depending on the context of the study and the nominal localisation accuracy expected. To improve understanding, the

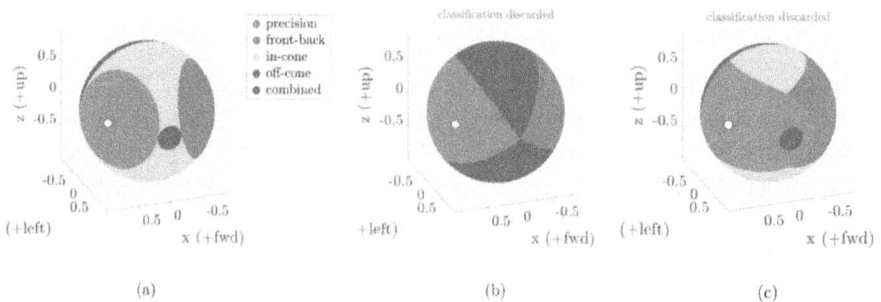

(a)   (b)   (c)

Figure 3.
Confusion type as a function of response position on the sphere, for a target at spherical coordinates (35°, 10°) and a listener facing X with his left ear pointing towards Y. (a) Proposed classification scheme, (b) classification used in Stitt et al. [10] based on polar angle only, and (c) attempt at solving pole compression issues of (b).

evolution of confusion zones for a 20° threshold and various target position is illustrated in **Figure 4**. The sum of the four confusion category rates always sums to 100%.

The distinction between in- and off-cone confusions is inspired from the duplex theory [36, 37], separating responses based on whether they are caused by misinterpreting monaural cues (in-cone confusions) or binaural cues (off-cone confusions). The commonly cited front-back confusion category has been maintained, despite not having a clearly identified origin in signal symmetry, as it represents a behaviour frequently observed in localisation studies [38]. Other confusion categories have been considered for this scheme, such as up-down or combined up-down-front-back confusions. They have been discarded however, as their representative patterns were not prevalent in the ≈10000 participant responses analysed in Section 4 or the meta analysis on ≈80000 responses in free field by Best et al. [38].

Compared to traditional cone-of-confusion classifications defined using only polar angle [8, 10, 16, 29], the main drawback of the proposed scheme is that it is susceptible to ITD mismatch. By only looking at the difference in *polar* angle between target and response, these classifications are not impacted by participants misinterpreting the ITD of the target, focusing on monaural cues interpretation characterisation. As illustrated in **Figure 3b**, the problem of these classifications is that they have high rate of false error detection at the poles of the interaural coordinate system, were a small shift in response can be interpreted as *e.g.* a front-back confusion instead of a precision error.

An attempt was made to propose a new scheme, inspired by the one used in Stitt et al. [10], alleviating the pole issue by increasing the (polar) spread of the precision zone as targets near the poles, constraining said spread to always span 45 of great-circle angle when projected on the sphere. As illustrated in **Figure 3c**, this constraint results in a undesirable warping of the precision error zone for targets within a certain lateral distance from the poles.

The solution proposed for studies needing a classification based on monaural cues interpretation alone is to extend the proposed scheme, artificially adjusting the lateral position of targets prior to the classification to discard errors related to ITD mismatch. This adjustment can be made on a per-participant/target basis, replacing the lateral angle of targets by the mean lateral angle of their associated responses prior to the classification. It can also be performed on a per-response basis by simply assuming that targets and responses always have the same lateral position. The case study of Section 4 uses the second, simple, non-adaptive form of the classification scheme.

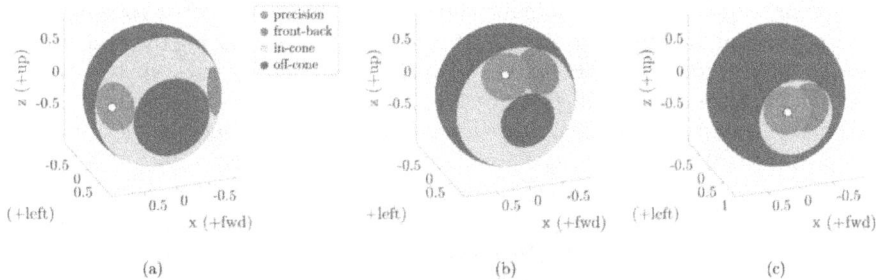

Figure 4.
Confusion type as a function of response position on the sphere for the proposed classification scheme with an angle threshold of 20 and a listener facing X with his left ear pointing towards Y. Target at spherical coordinates (a) (35°, 10°), (b) (70°, 40°), and (c) (80°, 10°).

3.1.5 Azimuth, elevation, lateral, and polar errors and biases

Lateral and polar errors are defined as the absolute difference between target and response positions in interaural coordinates. They are used to project localisation errors onto spatial dimensions associated with separate cues in the HRTF, allowing for an analysis of their independent contribution to the overall performance. Both are defined in [0°:180°], where smaller values correspond to better localisation performances. In the methodology, lateral and polar errors will be evaluated only on responses classified as *precision* confusions, hence referred to as *local* lateral and polar errors. This limitation allows to avoid the discontinuities discussed in Section 2.1.2 as well as the hazardous interpretation of values compounding local errors and spatial confusions.

As mentioned in Section 2.1.3, compression at the poles will lead to artificially inflated polar errors for targets near the interaural axis. A weight, proportional to the target lateral position, can be applied to the polar error to compensate for the compression, defining the *polar error weighted* as:

$$\text{polar\_error\_weighted} = \text{polar\_error} * cos\left(\alpha_{target}\right) \tag{4}$$

This weight is designed so that, for a target and a response that share the same lateral angle, the polar error weighted is equal to the arc length (great-circle) that separates them, regardless of said lateral angle. Note that while lateral error is not impacted by pole compression, it 'folds' near the interaural axis: random responses will overall have a lower local lateral error for targets in this region. This is a valuable feature of the interaural system when assessing the symmetric contribution of binaural cues (ITD/ILD) to localisation error. It can nonetheless lead to artificially deflated lateral errors when used in a different context.

Azimuth and elevation errors are defined as the absolute difference between target and response positions in spherical coordinates. They correspond to a more traditional projection of spherical coordinates, more intuitive yet no longer guided by auditory cue separation. Like interaural errors, azimuth and elevation errors are defined [0°:180°] and will be used only for local precision evaluation. As for polar error, azimuth error compression near the poles can be compensated for, defining the *azimuth error weighted* as:

$$\text{azimuth\_error\_weighted} = \text{azimuth\_error} * cos\left(\varphi_{target}\right) \tag{5}$$

In addition to absolute errors, *signed* lateral and elevation errors are used in the methodology. Mean signed errors, referred to as *biases*, are typically used to examine systematic rotational biases, induced for example by an offset between the tracking system used for measuring the HRTF and that used during the evaluation task, or reporting bias. As for absolute errors, usage of both metrics will be restricted to responses classified as precision confusions.

Finally, lateral and elevation *compression* errors are used to highlight space compression and dilation effects. *Lateral compression*, is defined as $\|\alpha_{target}\| - \|\alpha_{response}\|$, so that a positive error corresponds to a compression towards the median plane ZX. Respectively, a negative error corresponds to a dilation away from the median plane. Similarly, the *elevation compression* is defined as $\|\varphi_{target}\| - \|\varphi_{response}\|$, so that a positive error corresponds to a compression towards the horizontal plane XY. Respectively, a negative error corresponds to a dilation away from the horizontal plane. Compression errors are for example used to characterise a pointing bias caused by the reporting interface, or to detect lateral compressions resulting from an ITD mismatch between the presented HRTF and that of the participants.

3.1.6 Sphere regions

The decomposition of the analysis in sphere regions depends on the context. As such, there exists no one ideal decomposition scheme. To support the case study presented in the next section, the sphere will be split into 6 regions: *front-up* ($x > 0$ and $z > 0$), *front-down* ($x > 0$ and $z < 0$), *back-up* ($x < 0$ and $z > 0$), *back-down* ($x < 0$ and $z < 0$), *left* ($y > 0$), and *right* ($y < 0$). This scheme has been chosen to best highlight region specific behaviours while remaining manageable, based on a pre-liminary analysis of the experiments studied in Section 4. The redundant *left* and *right* regions have been added for systematic checks on lateralisation discrepancies in participant responses.

3.2 Methodology

The methodology is proposed as a set of analysis steps, each building on the previous one to provide a comprehensive assessment of participants localisation performance.

3.2.1 Evaluation task characterisation

The first step of the analysis is to assess how much of the space, *i.e.* sphere, has been tested during the localisation task. In addition to depicting the grid of tested positions, this step reports its space coverage statistics as defined in Section 3.1.2. This provides readers with a simple set of metrics that reflect the spatial thorough-ness of the evaluation, a value they can use to qualify the study's conclusions as well as for inter-study comparisons.

Atypical evaluation grids and their potential impact on participant results should also be discussed here. An evaluation on frontal field positions alone is likely to result in better overall performance compared to one encompassing the whole sphere, due to known variations of perceptual accuracy across sphere regions [5]. When using such grids, reporting metrics chance rates, *i.e.* their values for responses randomly distributed on the sphere, as proposed by Majdak et al. [14] can greatly help readers appreciate the presented results. Another problematic example is the use of evalua-tion grids sparse enough for participants to identify and recall the tested positions, likely impacting participants performance and associated conclusions.

Finally, the stimulus characteristics (type, duration, *etc.*) as well as the reporting method should be described and discussed here, so that any systematic bias they may have on participant responses can be detected during the analysis.

3.2.2 Assess global extent of localisation error

The objective here is to get a rough overview of participant performance during the localisation task, simply answering the question "how far were responses from the true target position?". The assessment is based on the great-circle error as defined in Section 3.1.3.

3.2.3 Assess critical localisation confusions

The next step consists in separating small precision errors from critical confu-sions. The nature and types of confusions is characterised early on as they can have a critical impact on localisation performance, often far more detrimental than local localisation accuracy issues. This characterisation is performed using one of the classification methods defined in Section 3.1.3.

3.2.4 Assess local extent of localisation error

This next step takes a closer look at responses classified as precision errors, *i.e.* the non-confused responses, to examine the local localisation performance. The mean great-circle error and angular direction of responses classified as precision confusions is computed to analyse the extent of local errors. Note that this metric does not depend on the confusion classification method used, as precision errors are defined using the same criterion in both methods. Conclusions drawn from this local analysis should naturally be leveraged by the percentage of responses it encompasses.

3.2.5 Horizontal and vertical decomposition of the localisation error

Whether or not this step should be included in the analysis, and which metrics it should make use of, depends on the context of the study. An experiment focusing on perceptual ITD adjustment for example would likely make use of both local lateral error as well as lateral compression. A training program attempting to fine tune participant interpretation of monaural cues would on the other hand base its evaluation on the local polar error. For some studies, this decomposition will not make sense and should be avoided to limit Type I error inflation.

3.2.6 Decompose the analysis across sphere regions

This final step consists in repeating all of the above, decomposing the analysis based on target positions to assess how participants fared in specific regions of the sphere. Given the loss of statistical power and the additional clutter that this analysis represents, it only needs to apply to those studies interested in characterising spatial imbalances in performance. The decomposition can then be performed using either a sphere splitting scheme as the one described in Section 3.1.6, or on a per-target position basis. For example, this approach can be used to support the design of HRTF learning programs that would focus dynamically on those regions/confusions that are the most problematic [9].

To further characterise local localisation behaviours, the analysis can be completed by evaluating average response positions and spherical response distributions. The former, computed by summing local great-circle error *vectors*, as discussed in Section 3.1.3, will help characterise variations of localisation accuracy across sphere regions [21]. The latter, characterised using Kent distributions (see Section 3.1.3), will provide the statistical framework to assess the significance of those variations.

4. Case study

The methodology defined in the previous section is applied here to build a comparative analysis on a selection of studies, focusing on the use of, and adaptation to, binaural cues for auditory localisation. The objective of this case study collection is not so much to present a thorough comparison of these studies as to illustrate how the methodology can be applied to a practical use case, and how its constituting metrics react to concrete scenarios. To further focus the case-study on these points, significance assessment is based on the overlapping of estimated distributions Confidence Intervals (CIs) rather than on null-hypothesis tests [39].

4.1 Study selection overview

Several studies of the impact of HRTF training on localisation accuracy have been selected from existing literature, for which authors graciously provided raw participant data used in the comparative analysis. A short description of each study is provided in the next section, reporting only those elements that concern the present analysis.

Common to most of the presented studies is the notion of HRTF *perceptual quality*. This term refers to the perceptual matching, localisation wise, between a participant and an HRTF. A low quality HRTF is one that results in bad localisation accuracy. Inversely, the higher the quality, the better the localisation accuracy, the highest quality match corresponding in theory to one's own HRTF. Replicating the potential outcomes of selecting an HRTF from an existing database, three degrees of perceptual matching are considered in these studies in addition to individual HRTF: *worst-match*, *random-match*, and *best-match* HRTF. Best and worst-match HRTFs represent respectively a best and worst case outcome, typically obtained by asking participants to perform a localisation task with, or a perceptual ranking of, an existing set of HRTFs.

4.1.1 Study description: *exp-majdak*

Majdak et al. [14], a 2010 study on the impact of various reporting methods during training with their individual HRTF. 10 participants trained on auditory localisation: 5 reporting perceived localisation positions with their hand, 5 with their head. Each participant completed 600–2200 localisation trials over a span of 2–32 d. Training and evaluation were performed within each trial: a session was composed of 50 trials, completed in 20–30 min. Each trial consisted of a localisation task with feedback, testing participants on 1380 positions overall, distributed on a sphere, using a 500 ms burst of white noise as stimulus. As the reporting method proved to have only a small impact on training efficiency, the 10 participants have hereafter been aggregated in a single group (**grp-majdak-indiv**), focusing the analysis on the impact of HRTF quality on performance evolution.

4.1.2 Study description: *exp-parseihian*

Parseihian and Katz [8], a 2012 study on accommodation to non-individual HRTF. 12 participants trained on auditory localisation, each completing 3 sessions of 12 min each on 3 consecutive days. Each session consisted of an interactive audio localisation game followed by a localisation task evaluation testing participants on 25 positions distributed on a sphere, using a 180 ms sequence of white noise bursts as stimulus. Before training, each participant ranked a set of 7 *perceptually orthogonal* HRTFs [40, 41] from the LISTEN database [42] based on localisation accuracy as perceived during predefined audio trajectories. The best and worst-match HRTF for each participant was extracted from this ranking. Participants were then divided into 3 groups: 2 that trained with their individual HRTF (**grp-parse-indiv**), 5 with the best-match HRTF (**grp-parse-best**), and 5 with the worst-match HRTF (**grp-parse-worst**). An additional 2 groups that performed only 1 training session are not considered in the current analysis. The ITDs of all HRTFs were adjusted based on individual participant head circumference, using a model derived from a regression between measured ITDs and morphological parameters. This technique is used as a practical method, easily carried out by end-users, to maximise initial localisation performance accuracy.

4.1.3 Study description: **exp-stitt**

Stitt et al. [10], a 2019 study on accommodation to non-individual HRTF. 16 participants trained on auditory localisation, each completing 10 sessions of 12 min each over a span of 10–20 weeks. The worst-match HRTF selection, training game, stimulus, and tested audio source positions during the localisation task evaluation at the end of each training session were the same as those of **exp-parseihian**. Participants were divided into 2 groups: 4 training with individual HRTFs (**grp-stitt-indiv**) and 8 with worst-match HRTFs (**grp-stitt-worst**). An additional 8 participants trained for only 4 sessions with their worst-match HRTFs are not considered in the current analysis.

4.1.4 Study description: **exp-steadman**

Steadman et al. [15], a 2019 study on accommodation to non-individual HRTF. 27 participants trained on auditory localisation, each completing 9 sessions of 12 min each over a span of 3 d. A localisation task evaluation was conducted at the beginning and end of each day as well as between each training session the first day, testing participants on 12 positions distributed on a sphere using a 1.6 s stimulus merging bursts of white noise and speech signal. All participants trained with the same randomly-matched HRTF selected from the 7 LISTEN database of **exp-parseihian**. Participants were distributed in 3 groups, training on various gamified and interactive versions of an audio localisation game, aggregated as one group in the current analysis (**grp-steadman-random**). An additional 9 participants, acting as a control group not undertaking training, are also not considered in the current analysis, as well as the results of a parallel evaluation task performed on another HRTF than that used during training.

4.1.5 Study description: **exp-poirier**

Poirier-Quinot and Katz [9], a 2021 study on accommodation to non-individual HRTF. 12 participants trained on auditory localisation (**grp-poirier-best**), each completing 3 sessions of 12 min each over a span of 3–5 d. Participants trained using a best-match HRTF selected from the 7 LISTEN database of **exp-parseihian**, though the simplified subjective selection method was only concerned with identifying the best-match HRTF. An additional 12 participants trained with their best-match HRTF in a reverberant condition are not considered in the current analysis. Each session consisted of an interactive audio localisation game followed by a localisation task evaluation testing 20 positions distributed on a sphere using the same stimulus as in **exp-parseihian**.

4.2 Application of the methodology

4.2.1 Time alignment of evaluation sessions

In all these experiments, the training sessions lasted for 12 min, except for **exp-majdak** where both training and evaluation were performed in a single block of 20–30 min. According to **exp-majdak**, the evaluation itself took half that time, leaving a per-session training duration equivalent to that of the other studies. A time realignment across experiments was executed such that the evaluation sessions compared are separated by equivalent training durations. Thus, the sessions have been renumbered to account for changes in protocol.

In the analysis, evaluation sessions are numbered from 1 to 11, each separated by a 12 min training. **Exp-poirier** and **exp-parseihian** only performed 3 training sessions, hence the missing data-points in subsequent figures. Likewise, **exp-stitt** and **exp-majdak** did not report pre-training performances, missing session 1 data-points. Finally, the number of evaluations in **exp-steadman** spreads out from session 4 onward, switching from an evaluation session after each training to an evaluation at the beginning and end of each 3-sessions training day.

4.2.2 Evaluation task characterisation

The space coverage of target positions evaluated during the localisation task of each study are reported in **Figure 5**. The high density of the grid of **exp-majdak** results in a very low average sc_{angle} compared to those of the other experiments. Its comparatively high standard deviation is due to the absence of test positions in the bottom part of the sphere (polar gap). For comparison, a homogeneous grid with the same number of points would have yielded $sc_{angle} = 0.5° \pm 0.003$. Distribution homogeneity is also responsible for the lower sc_{angle} standard deviation value observed in **exp-poirier** compared to that of **exp-parseihian** and **exp-stitt**. Finally, **exp-steadman**, with fewer test points and a polar gap in the bottom hemisphere, has the highest sc_{angle} value and standard deviation.

As could be expected, all the grids present high sc_{shape} values, being overall evenly distributed on the sphere. Grid density around polar gaps impacts the metric, explaining why **exp-poirier** value is higher than that of **exp-majdak** while both grids are evenly distributed: removing polar gap contributions in these grids would yield sc_{shape} values of 0.91 and 0.84 respectively.

Two different reporting methods were used in the five studies: head pointing (**exp-majdak** and **exp-steadman**) and hand pointing (**exp-majdak, exp-parseihian, exp-steadman, exp-poirier**). This should have little to no impact on the comparative analysis however, as both methods lead to similar reporting biases [32]. **exp-parseihian, exp-stitt**, and **exp-poirier** used the same stimulus: a 180 sequence of three white noise bursts. **Exp-majdak** used a slightly longer, unique burst of 500 ms, and **exp-steadman** used a 1.6 s stimulus composed of both white noise bursts and speech signal. All these stimuli are likely to present the transient energy and the broad frequency content necessary for auditory space discrimination [43, 44]. The difference in stimulus duration may have

$sc_{angle} = 0.5° \pm 1.2$
$sc_{shape} = 0.76 \pm 0.12$

$sc_{angle} = 28.8° \pm 16.5$
$sc_{shape} = 0.78 \pm 0.11$

$sc_{angle} = 36.0° \pm 8.7$
$sc_{shape} = 0.89 \pm 0.03$

$sc_{angle} = 60.0° \pm 15.3$
$sc_{shape} = 0.73 \pm 0.13$

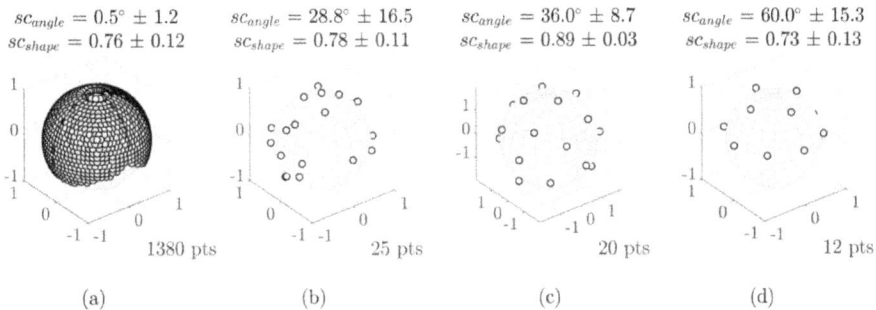

1380 pts   25 pts   20 pts   12 pts

(a)   (b)   (c)   (d)

Figure 5.
Space coverage statistics of the evaluation task in the selected studies (a) majdak, (b) parseihian/stitt, (c) poirier and (d) steadman.

repercussions in the analysis, as the participants can initiate more head movements to facilitate auditory localisation during the presentation of longer stimuli [45]. While adaptive rendering (*i.e.* dynamic cues) was disabled during stimulus presentation in **exp-parseihian, exp-stitt**, and **exp-poirier**, this is not explicitly stated in **exp-majdak** and **exp-steadman**.

4.2.3 Assessing the global extent of localisation error

The evolution of great-circle angle error across studies and training sessions is reported in **Figure 6**. Besides the clear benefit of training observed in all studies, the metric also highlights the overall positive impact of HRTF quality on initial performance. Interestingly, while the results from **exp-parseihian** suggest a similar intra-HRTF quality/performance relationship, it reports larger great-circle angle errors compared to those of the other experiments. This point already illustrates how differences in evaluation protocols or inter-participant variations may complicate the comparison of results across studies, as discussed in Section 4.3.

4.2.4 Assessing the critical localisation confusions

Much like the great-circle error, precision confusion rates can be used to assess performance evolution during training, as illustrated in **Figure 7**. Trends observed on initial precision rates and their evolution reflect the observation made on the great-circle error analysis. Precision rates and great-circle angle values are indeed highly correlated across training sessions, with correlation coefficients in $[-1.0:-0.9]$ for all studies. As each confusion rate aggregates all the responses of a participant during an evaluation session however, their CI is by construction often wide enough to confuse the analysis compared to that based on great-circle errors.

This widening of the CIs is particularly apparent in the comparison of the other confusion rates, reported in **Figure 8** for the evaluation that took place after the first training session. While a trend indeed suggests that the amount of confusions increases with decreasing HRTF quality, overlapping CIs often prevent any definite conclusion. Observing these rates can still help inform the analysis, as the poor performance of **grp-parse-indiv** on great-circle error observed in the previous section can be partly attributed to their high in-cone confusion rates, while their off-cone confusion rate is on par with that of **grp-stitt-indiv** and **grp-majdak-indiv**.

Maybe the most interesting use of confusion rates is to decompose the overall performance evolution. As illustrated by its confusion rate evolution in **Figure 9**, **grp-stitt-worst** performance evolution observed in **Figure 6** should, confusion wise, mainly be attributed to improvements in front-back confusions during training.

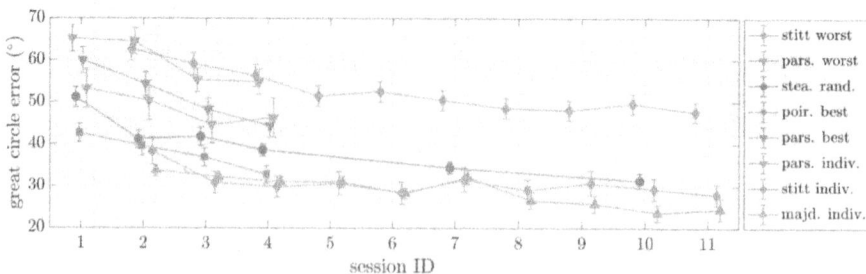

Figure 6.
Great-circle error mean and CI evolution across sessions and experiments. The great-circle error value for random responses is of 90° for all experiments.

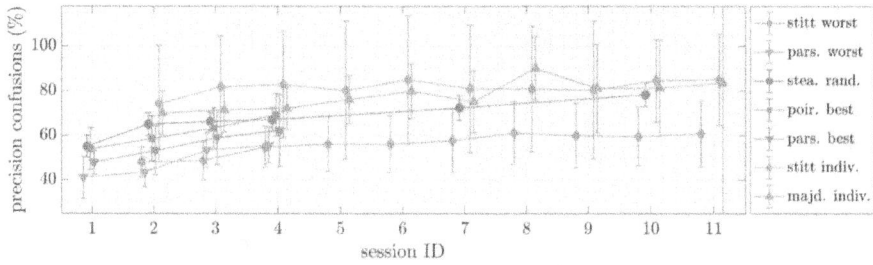

Figure 7.
*Precision confusion rates mean and CI evolution across sessions and experiments. **Grp-parse-indiv** was removed from the figure, composed of only 2 participants, resulting in a CI so large it confused the whole plot.*

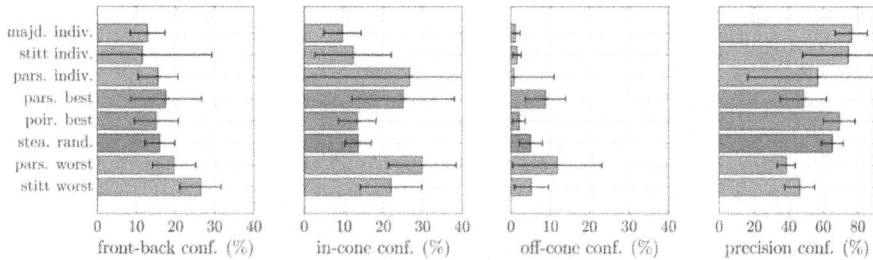

Figure 8.
Confusion rates after the first training session across experiments.

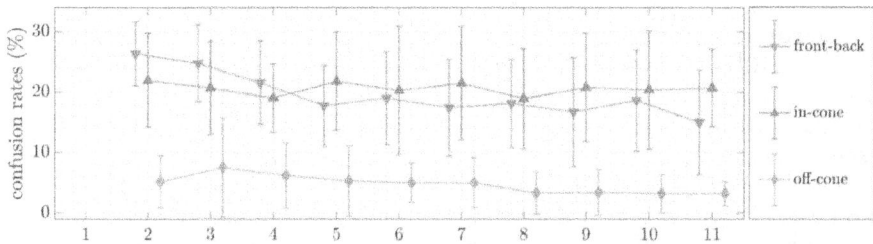

Figure 9.
*Confusion rates mean and CI evolution across sessions for **grp-stitt-worst**.*

4.2.5 Assessing the local extent of localisation error

Results of the confusion classification indicate that roughly 50% of responses were within the vicinity of the target (precision errors) after the first training session across experiments. The analysis here focuses on these responses, assessing local accuracy issues to complete that on localisation confusions.

Figure 10 reports local great-circle errors across training sessions and experiments. Looking once more at **grp-stitt-worst**, their local accuracy did not improve during training, oscillating around 25°. The improvement seen on overall great-circle error for that group can therefore be solely attributed to the reduction in front-back confusions reported in the previous section. Likewise, the 10° improvement on overall great-circle error observed for **grp-parse-worst** between sessions 2 and 3 can be attributed to a reduction in confusion rates, as it does not appear on local great-circle error. Separating the contribution of confusions from that of local accuracy also reveals a significant difference between **grp-stitt-indiv** and **grp-**

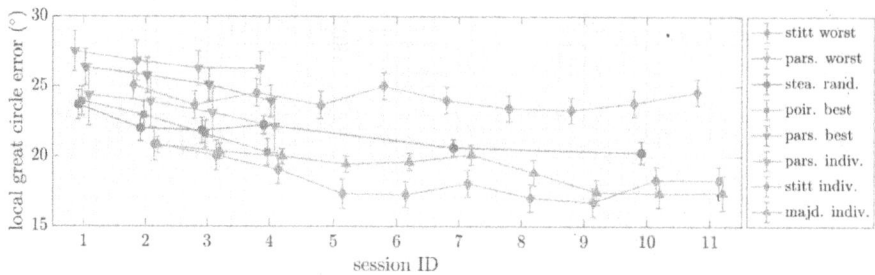

Figure 10.
Local great-circle error mean and CI evolution across sessions and experiments.

majdak-indiv improvement of local great-circle error between sessions 2 and 6, not visible on global great-circle error.

4.2.6 Horizontal and vertical decomposition of the localisation error

Local lateral error evolution across sessions for all experiments is reported in **Figure 11a**. As expected, initial performances indicate that participants using individual HRTF were quite apt at lateral localisation, accustomed as they were to the presented ITD and ILD cues. **Exp-poirier, exp-stitt,** and **exp-parseihian** used a similar ITD adjustment scheme, slightly improved in its last iteration for **exp-poirier** compared to that of **exp-stitt,** itself an incrementation on that of **exp-parseihian.** As such, the progression of initial lateral errors between **grp-parse-worst, grp-stitt-worst,** and **grp-poirier-best** can be expected. The performance of **grp-steadman-random,** on par with that of participants using ITD-adjusted or individual HRTFs, could be either attributed to the small number of evaluation positions (similar to that used during training), or to the 1.6 s burst and voice stimulus used as compared to the 180 ms to 500 ms burst trains used in the other experiments.

Participants trained with individual HRTF did not improve much on local lateral error overall, starting at ≈11° after the first training session and only improving to at ≈9° after the last. Comparison of performance evolution between groups training with a worst-match HRTF (**grp-parse-worst** and **grp-stitt-worst**) against that of groups training with a best-match HRTF (**grp-parse-best** and **grp-poirier-best**) suggests a positive impact of HRTF quality on potential local lateral error improvement. It would also seem that the ITD adjustment applied in **exp-parseihian** and **exp-stitt** was not sufficient to compensate for poor HRTF quality regarding lateral localisation accuracy.

Focusing on local lateral compression evolution, **Figure 11b** reveals a systematic over-estimation of the lateral angle across experiments, *i.e.* participants overall reported targets closer to the inter-aural axis poles than they truly were. Analysis of session 2, after the first training session, indicates that 62% of the 73 participants presented an overall lateral compression of less than −5°, against only 4% presenting one above 5°.

Local polar error evolution across sessions for all experiments is reported in **Figure 12a**. Overall performance was still a function of HRTF quality, but for **grp-parse-indiv** poor performance prior to training and **grp-steadman-random,** on par with **exp-stitt** and **exp-majdak** control groups using individual HRTFs. The impact of training is hardly more pronounced than that observed on local lateral error. Training still helped lower local polar error overall, with even participants using individual HRTFs slightly improving during training: **grp-stitt-indiv** and **grp-**

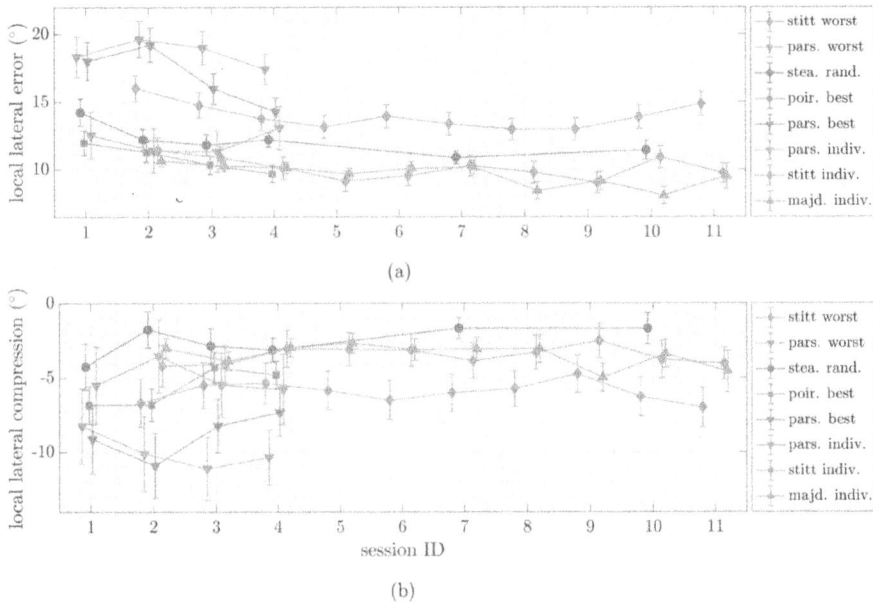

Figure 11.
(a) local lateral error, and (b) local lateral compression evolution across sessions and experiments.

majdak-indiv gained ≈3° in local polar accuracy over the course of training, roughly identical to the improvement observed on local lateral accuracy. Note here that an analysis based on the overall polar error, *i.e.* taking into account confusions, would have suggested ≈12° improvement after training for these two groups. Finally, most of the improvement on local polar error occurred during the early stage of the training, decreasing of ≈7° between sessions 1 and 2 in average over all experiments, not considering **exp-stitt** and **exp-majdak** as participants were not tested prior to training, and of only ≈7° between sessions 2 and 4.

The analysis of local elevation compression also reveals a stronger tendency to under-estimate target elevation, *i.e.* responses closer to the horizontal plane than the true target, than that observed on local lateral compression. Across experiments, 38% of the 73 participants presented a local elevation compression of more than 5° after the first training session, compared to 14% for elevation dilation. A trend suggests that local elevation compression is quickly corrected during the first train- ing session and remains at a relatively constant value regardless of the method or number of training sessions. The surprisingly high plateau reached by **grp-majdak- indiv** compared to **grp-stitt-indiv**, also training on individual HRTFs, could be attributed to the the difference in tested grid positions: **exp-majdak** presented far more targets near the 90° elevation pole than **exp-stitt**.

4.2.7 Decompose the analysis across sphere regions

This section illustrates how splitting results analysis across sphere regions might highlight spatial imbalances in performance. To avoid further cluttering the chap- ter, only two example decompositions will be presented: confusion rates based on sphere regions, and local great-circle error based on individual target locations.

Decomposition of confusion rates based on the regions defined in Section 3.1.6 is illustrated **Figure 13**. Results displayed are aggregated over all five studies, to focus the analysis on general binaural localisation behaviours. The first noticeable result is

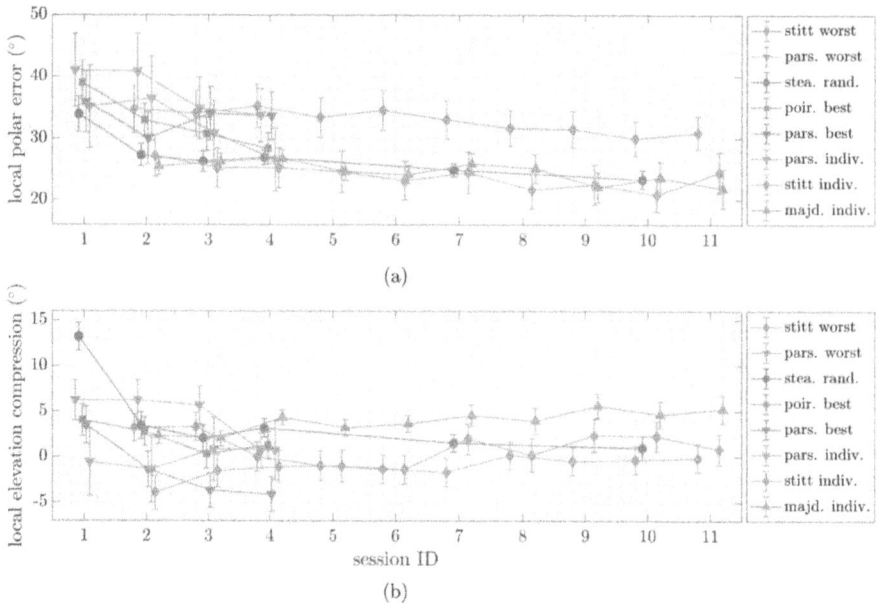

Figure 12.
Participants (a) local polar error, and (b) local elevation compression across training sessions and experiments.

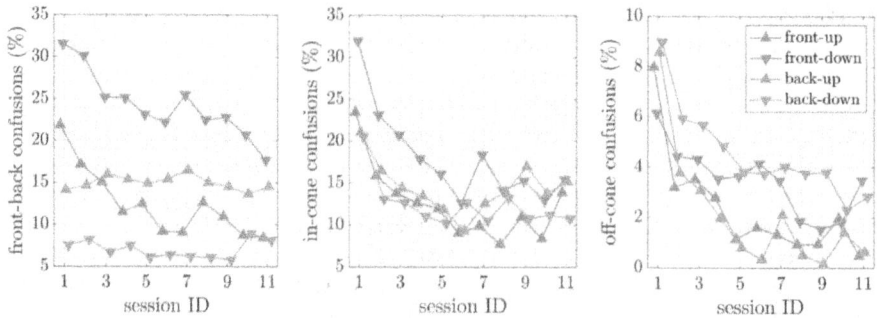

Figure 13.
Evolution of confusion rates across sessions, decomposed based on sphere regions, aggregated over all experiments.

that targets in the front-down region were the most susceptible to front-back and in-cone confusions initially, resulting in a very low precision rate (30% vs. 47% and more for the other regions) prior to the first training session. Interestingly, confusion rates in the front-down region were systematically higher than those in the front-up region, for all but off-cone confusions. The initial rate of front-back confusions of targets in front of participants, more than twice that of targets behind them, is likely due to the absence of visual feedback during the localisation task, increasing likelihood of perceiving a sound as behind if they cannot see its source, regardless of HRTF cues.

A second interesting result is the negligible evolution of front-back confusions for targets in the back regions throughout training (*i.e.* back-to-front). While the precision rate of all regions increased, and front-back confusions dropped for front regions, training seemed to have no impact on front-back rates in the back region. Analysis of per-region accuracy however revealed that the local great-circle error decreased evenly across regions, from ≈25° in session 1 to ≈21° in session 11.

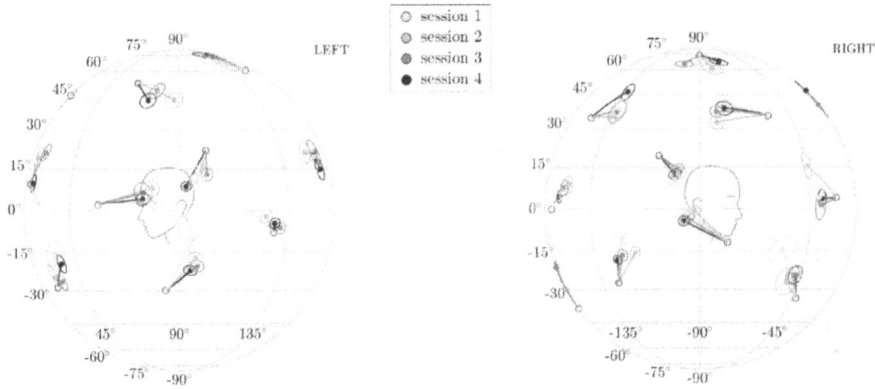

Figure 14.
*Evolution of mean response locations across targets and sessions in **exp-poirier**. Hollow circles represent target positions. Filled circles represent mean response locations, surrounded by standard error ellipses computed using Kent distributions.*

These observations suggest that future training programs could be improved by focusing slightly more on reducing front-back and in-cone confusions in the front-down region. Stagnating rates, such as that of front-back confusions in the back-up region, around 15% across sessions, would also suggest that there is room for improvement in the design of didactic training programs that would aid participants towards reaching 0% confusion rates.

Further refining the analysis, **Figure 14** focuses on the assessment of mean response locations for each target presented in **exp-poirier**. Mean response locations were obtained by summing local great-circle error vectors as discussed in Section 3.2.6. Their positions relative to targets, and the evolution of these positions during training, provides a thorough characterisation of participant's local accuracy evolution on the sphere. Additionally, the lateral and elevation compression effects observed in Section 4.2.6 are clearly visible, where mean responses are generally biased towards the interaural axes and/or the horizontal plane.

4.2.8 Handling initial performance offsets

This additional step in the analysis can be seen as an extension of the evaluation task characterisation proposed in Section 4.2.2 specific to the assessment of localisation performance *evolution*. It presents some of the techniques that exist to compare said evolution despite unbalanced initial conditions across studies or groups of participants.

Techniques have been proposed to conduct training efficiency analysis on unbalanced initial conditions. Stitt et al. [10] for example applied per-participant arithmetic normalisation, based on group baseline performances. Realigning initial conditions, this technique allows to focus the analysis on relative improvement, as illustrated in **Figure 15**.

Another technique for relative improvement comparison, used for example by Majdak et al. [31] and Poirier-Quinot and Katz [9], is to compare the coefficients of a regression applied on performance evolution. As mentioned in Section 2.3.2, two main regression models have been adopted to fit said evolution depending on the training stages represented in the data. **Figure 16** illustrates how both can be fitted to local great-circle error evolution across experiments. Groups performance evolution was first fitted to the exponential form in **Figure 16a**, resorting to the linear

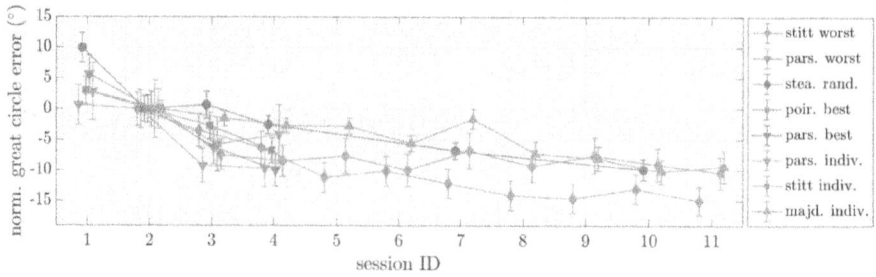

Figure 15.
Great-circle error evolution across sessions and experiments. Data normalised (subtraction) with group mean results of session 2 as reference.

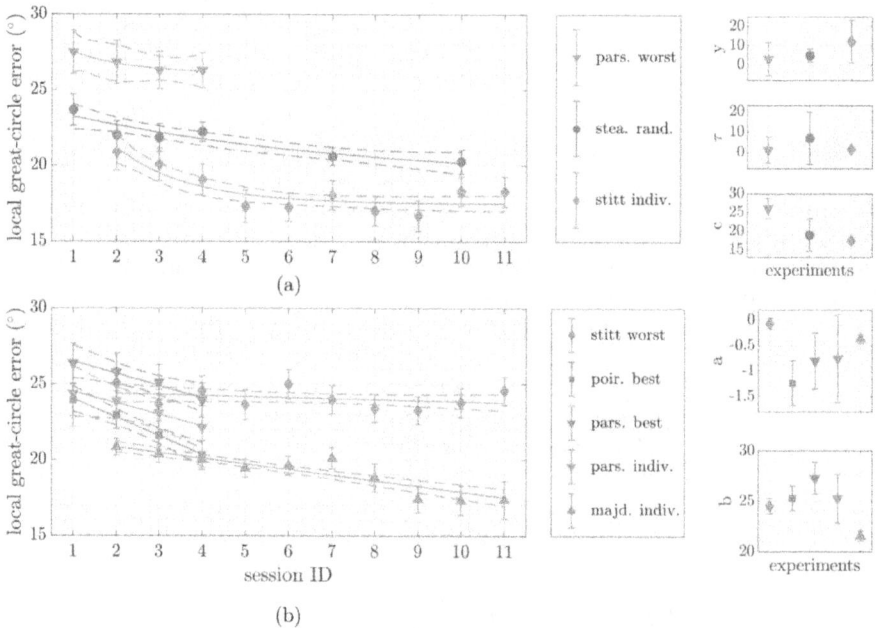

Figure 16.
Regressions on local great-circle error evolution across training and experiments, (a) exponential regression "$y_0 \times exp\left(-session_{ID}/\tau\right) + c$", and (b) linear regression "$a \times session_{ID} + b$". y_0 represents the initial performance, τ the improvement time constant, and c the long term performance. b represents the initial performance, a the improvement rate.

form in (b) when the evolution did not follow an exponential form, resulting in regression parameters CIs so wide as to prevent any meaningful interpretation. The use of a regression is particularly attractive, as it reduces the performance evolution analysis to a simple high level coefficient comparison, coefficients that can usually be interpreted in simple terms such as initial performance or improvement rate.

As mentioned, these techniques are generally applied to compensate for unbalanced initial performance. Although they are perfectly valid to assess the impact of HRTF quality or training efficiency on *relative* improvement, the scope of any conclusion made using them is greatly limited as the potential improvement margin naturally depends on initial performance.

4.3 Discussion

As illustrated throughout Section 4.2, drawing clear cut conclusions from the comparison of results from several studies is difficult at best. Most of the time, it is simply impossible, generally because of uncontrolled variations across test conditions. These variations, limiting both intra- and inter-study analysis, are discussed in this section.

4.3.1 Evaluation task

Variations in the evaluation protocols and procedures between studies in the literature present a challenge for comparing the multiple experiments. Different experimental design choices, such as reporting method, spectral content and duration of the stimulus, and evaluation grid, have a direct impact on the baseline performance of participants [32]. For example, given the choice by **exp-steadman** to use a random-match HRTF, the notable results of **grp-steadman-random** compared to those of the other groups could be attributed to the training program. However, the 1.6 sec stimulus (that may have enabled the use of head movements during the evaluation) may also have contributed to the improved performance of **grp-steadman-random** compared to the other studies that used 180 or 500 ms bursts [46].

The use of a unique grid for localisation tasks across studies would assuredly simplify results comparisons. Said grid could, for example, be designed to be homogeneously distributed on the sphere [35]. For more flexible test conditions, a series of test grids of increasing point densities could be defined, where test positions of any given grid would be present on its higher density neighbours, easing down-sampling for comparison. Regarding the stimulus used or the reporting method, a simple solution would be to settle on those that respectively optimise localisation accuracy [47] and minimise reporting bias [32]. Pending the adoption of common practices, the bias induced by those design choices could technically be assessed from the results of a control group using individual HRTFs.

Another issue when comparing performance evolution across studies is the alignment of the evaluation sessions for fair comparison. As proposed in Section 4.2.1, a simple solution is to align them based on training duration. Time alignment would seem a better option than its alternative, based on the number of positions presented during the training. Time is of direct interest for end-users, and an alignment based on presented positions would bias the analysis in favour of slower exploratory training paradigms.

Finally, the merging of both evaluation and training sessions, as used in **exp-majdak**, is not ideal in the context of inter-study comparison. Although this practice allows for a more granular analysis of performance evolution, it systematically leads to confusing analysis compared to studies alternating between training and evaluation sessions. Additionally, it would seem that the alternating design imposes a lesser constraint on the training paradigm itself, allowing for implicit learning strategies not focused on target localisation [48].

4.3.2 Intra- and inter-participant variations

Variations between participants' performance is an issue common to most psychophysical studies studies. Two aspects of these variations can become critical in the context of HRTF learning studies.

The first aspect concerns imbalances in initial participant performance across tested conditions. As discussed in Section 4.2.8, such imbalance is likely to weaken or void conclusions resulting from the analysis. For within experiment comparisons, a simple solution is to run a pre-training evaluation session, to then create groups of equivalent performance based on the metrics used in the analysis. The problem naturally worsens when dealing with inter-study analysis. The use of a control group using individual HRTF is again advised to serve as a baseline reference for the comparative analysis.

The second aspect concerns the difference in participants' immediate sensitivity to HRTF quality, and their ability to adapt to a non-individual HRTF. Both have been discussed in previous studies, where some participants were more prone to instantly benefit from a best-match HRTF [49] or to adapt to a poorly matched HRTF [10]. To avoid missing out on interesting behaviours due to the variance introduced by some participants, it is recommended to conduct a second pass of the analysis on sub-groups, for example aggregated based on their improvement rate [10]. Although the conclusions from the sub-group analysis may be weaker compared to an overall analysis, the technique provides readers with a more thorough understanding of the training as well as the potential advantages and limitations of the tested conditions.

4.3.3 Procedural versus perceptual learning

In the present context, procedural learning refers to participants becoming familiar with the various aspects of the localisation task, resulting in a performance improvement that is not due to an accommodation to HRTF specific cues (perceptual learning). As of yet, there exists no model for *a posteriori* dissociating the contribution of both types of learning to performance evolution. Intra-study comparisons would most likely not be affected since one could generally assume that the procedural learning has a similar impact on all tested conditions. However, by not allowing the procedural learning to plateau before the first evaluation, the generalisation of a study conclusions become problematic when one needs to compare the results from various studies based on different protocols.

Results of control groups generally prove extremely valuable during inter-study comparison. Participants only taking part in the evaluation and not the training, as in **exp-steadman**, can provide a good insight on the impact of the evaluation task implementation on performance across experiments. Even better, the inclusion of a control group using their own HRTF, as in **exp-stitt** and **exp-parseihian**, provides a solid baseline to dissociate procedural from perceptual learning during both intra- and inter-study analysis.

Additionally, simple experimental design choices can be applied to avoid having to deal with certain forms of procedural training. The proprioceptive adjustment required for accurately reporting perceived positions [14] can for example be greatly accelerated by using a natural 3D reporting method coupled to a visual pointer [9], as well as providing a reference grid to help orientation in the sphere [31]. Thorough beta testing can further eliminate design flaws that participants can exploit to improve their performance, such as the use of too small a set of test positions, or unconstrained tracking allowing for small head movements during the stimulus presentation phase of the localisation task.

Other aspects of procedural training, such as having participants focus on the listening task, can only be removed by introducing a pre-experimental training session. Such a session was applied in **exp-majdak**, where participants trained for approximately 30 min on a localisation task coupling visual feedback and stereo panning. This pre-experimental training likely contributed to the smooth improvement in great-circle error by **grp-majdak-indiv** from session 2 onward compared to the disjointed improvement observed for **grp-stitt-indiv** between sessions 2 and 3 in **Figure 6**. Paradoxically, the only limitation of the pre-training proposed in **exp-majdak**, which did not use actual binaural signals, is that it does not familiarise participants with binaural rendering. Pending formal evidence, one may assume that there exists an adaptation process during which participants will grow consistent in their localisation estimation, even in the absence of feedback, much like the effect observed on HRTF quality ratings reported by Andreopoulou and Katz [50]. Regardless of whether this adaptation should be labelled as perceptual or procedural training, it will still interfere with the evaluation of training efficiency itself.

Overall, it is reasonable to assume that one could design a pre-training session that accommodates procedural learning in roughly 15 min, even taking into account this last point, and relaxing the time constraint imposed in **exp-majdak**. This session however still takes a non-negligible amount of time, which will contribute to participant fatigue and loss of focus. Because of this, it is likely that most experimental designs will continue to include aspects of procedural learning as a shared effect, equally impacting all tested conditions. An alternative solution would be to conduct a set of studies to measure and model the various aspects of procedural learning in the present context, so that its contribution to performance evolution could be dissociated from that of perceptual improvement even in the absence of a pre-training session.

5. Conclusion

This chapter presented a methodology for the assessment of auditory localisation accuracy in the context of HRTF selection and learning tasks. Based on existing metrics and decomposition schemes, the methodology consists of a series of steps guiding analysis towards the creation of comprehensive and repeatable performance assessments. A collected case-study was then proposed that compared the results of five contemporary experiments on HRTF learning and illustrates how the methodology can be applied to better understand participant performances and their evolution.

The initial intent of this chapter was to propose a set of metrics and an analysis workflow that would be adopted and adapted by the community to standardise the evaluation of localisation performance. In time, the standardisation would help simplify the comparison of results from different studies, allowing to assess hypotheses and draw conclusions beyond the scope of the constituting studies. While the proposed case-study provides a glimpse at the benefits of such standardisation, it is limited by one of, if not the most, major issue of inter-study comparison: the lack of a reference between tested conditions. Without this reference, conclusions drawn from the analysis can hardly be generalised, much like those that would result from a comparison between language learning techniques without *a priori* knowledge of participants learning abilities, or how different is the language learnt compared to their mother tongue.

As of now, the only applicable solution to provide such reference across studies is to systematically add a control group composed of participants using their own HRTF to the experiment. A large enough group composed of experts and novices alike would indeed provide a stable reference that can be used to assert a certain equivalence in *e.g.* the evaluation task before proceeding to inter-study performance comparison. However, this solution is rarely practical due to the complexity of the HRTF measurement process, which is the main incentive for HRTF learning in the first place. A somewhat less constraining, yet highly unlikely, scenario would be the creation and adoption of a unique evaluation platform, shared across all studies to formalise future HRTF selection methods and training program comparisons.

With luck, the issue will solve itself as the next generation of HRTF individualisation techniques render selection and training obsolete. In the meantime, methodologies such as the one proposed here should help improve the rigour of studies and consequently the understanding of the fundamental issues regarding auditory localisation and spatial hearing accommodation to non-individual HRTFs and their applications.

Acknowledgements

This work was funded in part through a fundamental research collaboration partnership between Sorbonne Université, CNRS, Institut ∂' Alembert and Facebook Reality Labs. This work was funded in part by the RASPUTIN project (ANR-18-CE38-0004, https://rasputin.lam.jussieu.frrasputin.lam.jussieu.fr) and an associated "Innov'up Faisabilité" grant from the Région Île de France. Portions of this work have been carried out in the context of the Sonicom project, that has received funding from the European Union's Horizon 2020 research and innovation programme under grant agreement No. 101017743.

Author details

David Poirier-Quinot[1†], Martin S. Lawless[2†], Peter Stitt[2†] and Brian F.G. Katz[2*†]

1 Sciences et Technologies de la Musique et du Son (STMS)—IRCAM, CNRS, Sorbonne Université, Paris, France

2 Sorbonne Université, Paris, France

*Address all correspondence to: brian.katz@sorbonne-universite.fr

† The listed authors contributed equally.

IntechOpen

References

[1] Letowski T, Letowski S. Localization error: Accuracy and precision of auditory localization. In: Advances in Sound Localization, Chapter 4. London: IntechOpen; 2011. DOI: 10.5772/15652

[2] Morimoto M, Aokata H. Localization cues of sound sources in the upper hemisphere. The Journal of the Acoustical Society of America. 1984; 5(3):165-173. DOI: 10.1250/ast.5.165

[3] Blauert J. Spatial Hearing: The Psychophysics of Human Sound Localization. Cambridge, Massachusetts, United States: MIT Press; 1997

[4] Katz B, Nicol R. Sensory Evaluation of Sound, Chapter Binaural Spatial Reproduction. Boca Raton: CRC Press; 2019. pp. 349-388

[5] Makous JC, Middlebrooks JC. Two-dimensional sound localization by human listeners. The Journal of the Acoustical Society of America. 1990; 87(5):2188-2200

[6] Kistler DJ, Wightman FL. A model of head-related transfer functions based on principal components analysis and minimum-phase reconstruction. The Journal of the Acoustical Society of America. 1992;91(3):1637-1647

[7] Bouchara T, Bara T-G, Weiss P-L, Guilbert A. Influence of vision on short-term sound localization training with non-individualized HRTF. In: EAA Spatial Audio Signal Processing Symposium. London: IEEE; 2019. pp. 55-60. DOI: 10.25836/sasp.2019.04

[8] Parseihian G, Katz BFG. Rapid head-related transfer function adaptation using a virtual auditory environment. The Journal of the Acoustical Society of America. 2012;131(4):2948-2957. DOI: 10.1121/1.3687448

[9] Poirier-Quinot D, Katz BF. On the improvement of accommodation to non-individual HRTFs via VR active learning and inclusion of a 3D room response. Acta Acustica. 2021; 5(25):1-17. DOI: 10.1051/aacus/2021019

[10] Stitt P, Picinali L, Katz BFG. Auditory accommodation to poorly matched non-individual spectral localization cues through active learning. Scientific Reports. 2019;9(1): 1063:1-1063:106314. DOI: 10.1038/s41598-018-37873-0

[11] Heffner HE, Heffner RS. The sound-localization ability of cats. Journal of Neurophysiology. 2005;94(5): 3653-3655. DOI: 10.1152/jn.00720.2005

[12] Middlebrooks JC. Virtual localization improved by scaling nonindividualized external-ear transfer functions in frequency. The Journal of the Acoustical Society of America. 1999; 106(3):1493-1510

[13] Carlile S, Balachandar K, Kelly H. Accommodating to new ears: The effects of sensory and sensory-motor feedback. The Journal of the Acoustical Society of America. 2014; 135(4):2002-2011. DOI: 10.1121/1.4868369

[14] Majdak P, Goupell MJ, Laback B. 3-D localization of virtual sound sources: Effects of visual environment, pointing method, and training. Attention, Perception, & Psychophysics. 2010; 72(2):454-469. DOI: 10.3758/APP.72.2.454

[15] Steadman MA, Kim C, Lestang J-H, Goodman DF, Picinali L. Short-term effects of sound localization training in virtual reality. Scientific Reports. 2019; 9(1):1-17. DOI: 10.1038/s41598-019-54811-w

[16] Zagala F, Noisternig M, Katz BFG. Comparison of direct and indirect perceptual head-related transfer function selection methods. The Journal of the Acoustical Society of America. 2020;**147**(5):3376-3389. DOI: 10.1121/10.0001183

[17] Leong P, Carlile S. Methods for spherical data analysis and visualization. Journal of Neuroscience Methods. 1998; **80**(2):191-200

[18] Wightman FL, Kistler DJ. Headphone simulation of free-field listening. II: Psychophysical validation. The Journal of the Acoustical Society of America. 1989;**85**(2):868-878

[19] Edmondson-Jones AM, Irving S, Moore DR, Hall DA. Planar localisation analyses: A novel application of a Centre of mass approach. Hearing Research. 2010;**267**(1–2):4-11

[20] Irving S, Moore DR. Training sound localization in normal hearing listeners with and without a unilateral ear plug. Hearing Research. 2011;**280**(1–2): 100-108

[21] Carlile S, Leong P, Hyams S. The nature and distribution of errors in sound localization by human listeners. Hearing Research. 1997;**114**:179-196. DOI: 10.1016/S0378-5955(97)00161-5

[22] Jammalamadaka SR, Sengupta A. Topics in Circular Statistics. Vol. 5. Singapore: World Scientific; 2001. DOI: 10.5772/15652

[23] Hofman PM, Van Riswick JG, Van Opstal AJ. Relearning sound localization with new ears. Nature Neuroscience. 1998;**1**(5):417-421. DOI: 10.1038/1633

[24] Trapeau R, Aubrais V, Schönwiesner M. Fast and persistent adaptation to new spectral cues for sound localization suggests a many-to-one mapping mechanism. The Journal of the Acoustical Society of America. 2016;

140(2):879-890. DOI: 10.1121/1.4960568

[25] Van Wanrooij MM, Van Opstal AJ. Relearning sound localization with a new ear. The Journal of Neuroscience. 2005;**25**(22):5413-5424. DOI: 10.1523/JNEUROSCI.0850-05.2005

[26] Wenzel EM, Arruda M, Kistler DJ, Wightman FL. Localization using nonindividualized head-related transfer functions. The Journal of the Acoustical Society of America. 1993;**94**(1):111-123. DOI: 10.1121/1.407089

[27] Zahorik P, Bangayan P, Sundareswaran V, Wang K, Tam C. Perceptual recalibration in human sound localization: Learning to remediate front-back reversals. The Journal of the Acoustical Society of America. 2006;**120**(1):343-359. DOI: 10.1121/1.2208429

[28] Honda A, Shibata H, Gyoba J, Saitou K, Iwaya Y, Suzuki Y. Transfer effects on sound localization performances from playing a virtual three-dimensional auditory game. Applied Acoustics. 2007;**68**(8): 885-896. DOI: 10.1016/j.apacoust.2006.08.007

[29] Martin RL, McAnally KI, Senova MA. Free-field equivalent localization of virtual audio. Journal of the Audio Engineering Society. 2001; **49**(1/2):14-22

[30] Yamagishi D, Ozawa K. Effects of timbre on learning to remediate sound localization in the horizontal plane. In: Principles and Applications of Spatial Hearing. Singapore: World Scientific; 2011. pp. 61-70

[31] Majdak P, Walder T, Laback B. Effect of long-term training on sound localization performance with spectrally warped and band-limited head-related transfer functions. The Journal of the

Acoustical Society of America. 2013; **134**(3):2148-2159. DOI: 10.1121/1.4816543

[32] Bahu H, Carpentier T, Noisternig M, Warusfel O. Comparison of different egocentric pointing methods for 3D sound localization experiments. Acta Acustica. 2016;**102**(1):107-118. DOI: 10.3813/AAA.918928

[33] Klein F, Werner S. Auditory adaptation to non-individual HRTF cues in binaural audio reproduction. Journal of the Audio Engineering Society. 2016; **64**(1/2):45-54

[34] Van Oosterom A, Strackee J. The solid angle of a plane triangle. IEEE Transactions on Biomedical Engineering. 1983;**2**:125-126. DOI: 10.1109/TBME.1983.325207

[35] Saff EB, Kuijlaars AB. Distributing many points on a sphere. The Mathematical Intelligencer. 1997;**19**(1): 5-11

[36] Middlebrooks JC, Green DM. Sound localization by human listeners. Annual Review of Psychology. 1991;**42**(1): 135-159

[37] Rayleigh L. XII. On our perception of sound direction. The London, Edinburgh, and Dublin Philosophical Magazine and Journal of Science. 1907; **13**(74):214-232. DOI: 10.1080/14786440709463595

[38] Best V, Brungart D, Carlile S, Jin C, Macpherson E, Martin R, et al. A meta-analysis of localization errors made in the anechoic free field. In: Principles and Applications of Spatial Hearing. Singapore: World Scientific; 2011. pp. 14-23

[39] Cumming G. The new statistics: Why and how. Psychological Science. 2014;**25**(1):7-29. DOI: 10.1177/0956797613504966

[40] Andreopoulou A, Katz BFG. Subjective HRTF evaluations for obtaining global similarity metrics of assessors and assessees. Journal of Multimodal User Interfaces. 2016b; **10**(3):259-271. DOI: 10.1007/s12193-016-0214-y

[41] Katz BFG, Parseihian G. Perceptually based head-related transfer function database optimization. The Journal of the Acoustical Society of America. 2012;**131**(2):99-105. DOI: 10.1121/1.3672641

[42] Warusfel O. IRCAM Listen HRTF Database. 2003. Available from: http://recherche.ircam.fr/equipes/salles/listen [Accessed: Sepember 29, 2018]

[43] Dramas F, Katz BFG, Jouffrais C. Auditory-guided reaching movements in the peripersonal frontal space. The Journal of the Acoustical Society of America. 2008;**123**(5):3723-3723. DOI: 10.1121/1.2935195

[44] Kumpik DP, Kacelnik O, King AJ. Adaptive reweighting of auditory localization cues in response to chronic unilateral earplugging in humans. The Journal of Neuroscience. 2010;**30**(14): 4883-4894. DOI: 10.1523/JNEUROSCI.5488-09.2010

[45] Woodworth RS, Schlosberg H. Experimental Psychology. Rev. ed. Oxford, England: Holt; 1954

[46] Wallach H. The role of head movements and vestibular and visual cues in sound localization. Journal of Experimental Psychology. 1940;**27**(4): 339-368

[47] Begault DR, Wenzel EM, Anderson MR. Direct comparison of the impact of head tracking, reverberation, and individualized head-related transfer functions on the spatial perception of a virtual speech source. Journal of the Audio Engineering Society. 2001; **49**(10):904-916

[48] Mendonça C. A review on auditory space adaptations to altered head-related cues. Frontiers in Neuroscience. 2014;**8**(219):1-14. DOI: 10.3389/fnins.2014.00219

[49] Poirier-Quinot D, Katz BFG. Assessing the impact of head-related transfer function individualization on performance: Case of a virtual reality shooter game. The Journal of the Audio Engineering Society. 2020;**68**(4): 248-260. DOI: 10.17743/jaes.2020.0004

[50] Andreopoulou A, Katz B. Investigation on subjective HRTF rating repeatability. In: Audio Engineering Society Convention. Vol. 140. New York, United States: Audio Engineering Society; 2016. pp. 9597:1-9597:959710

Chapter 4

The Influences of Hearing and Vision on Egocentric Distance and Room Size Perception under Rich-Cue Conditions

Hans-Joachim Maempel and Michael Horn

Abstract

Artistic renditions are mediated by the performance rooms in which they are staged. The perceived egocentric distance to the artists and the perceived room size are relevant features in this regard. The influences of both the presence and the properties of acoustic and visual environments on these features were investigated. Recordings of music and a speech performance were integrated into direct renderings of six rooms by applying dynamic binaural synthesis and chroma-key compositing. By the use of a linearized extraaural headset and a semi-panoramic stereoscopic projection, the auralized, visualized, and auralized-visualized spatial scenes were presented to test participants who were asked to estimate the egocentric distance and the room size. The mean estimates differed between the acoustic and the visual as well as between the acoustic-visual and the combined single-domain conditions. Geometric estimations in performance rooms relied upon nine-tenths on the visual, and one-tenth on the acoustic properties of the virtualized spatial scenes, but negligibly on their interaction. Structural and material properties of rooms may also influence auditory-visual distance perception.

Keywords: auditory-visual perception, virtual reality, egocentric distance, room size, performance room, concert hall, music, speech

1. Introduction

1.1 Desideratum

The multimodal perception, integration and mental reconstruction of the physical world provide us, amongst other things, with various modality-specific and modality-unspecific features such as colors, timbres, smells, vibrations, locations, dimensions, materials, and aesthetic impressions, which are or can be related to perceived objects and environments. A fundamental issue is the extent to which such features rely on the different modalities and their cooperation. The present study examined and experimentally dissociated the important modalities of hearing and vision by separately providing and manipulating the respectively perceivable information about the physical world, i.e., auralized and visualized spatial scenes. In everyday life, both the egocentric distance to visible sound sources and the size of a surrounding room are important perceptual features, since they contribute to spatial notion and orientation. They are also relevant about artistic renditions and performance rooms, as they relate, for instance, to the concept of auditory intimacy, an important aspect of the quality of concert halls [1–3]. Accordingly, both the perceived egocentric distance and the perceived room size were investigated, primarily in the context of artistic renditions.

1.2 State of the art

The interaction between hearing and vision occurs in the perception of various features, pertaining for example to intensity [4], localization [5–7], motion [8–10], event time [11, 12], synchrony [13], perceptual phonetics [14], quality rating [15], and room perception [16–18]. Regarding auditory-visual localization and spatial perception, research has focused mainly on horizontal directional localization to date, followed by distance localization, while room size perception has rarely been investigated. Two superior research objectives may be identified in the literature: One objective is the description of human perceptual performance and its dependence on physical cues. Within this context, distance perception was mainly investigated about its *accuracy*, and specifically via the experimental variation of the cues about the equivalent physical distance. The consideration of interfering factors such as the completeness and the integrity of the cues may be subsumed under this objective, too. Another objective is the modeling of internal processes of multisensory integration, which are closely related to the binding problem. The binding problem asks, how different sensory information is identified as belonging to the same event, object or stream, and thus is unified. According to Treisman there are "at least seven different types of binding": property, part, range, hierarchical, conditional, temporal, and location binding ([19], p. 171).

Experimental stimuli may be real objects (e.g., humans, loudspeakers, mechanical apparatuses) that have diverse physical properties and may bear meaning. Otherwise, the investigation of detailed internal mechanisms using behavioral experiments often calls for neutral objects or energetic events with a maximally reduced number of properties and without meaning (e.g., lights, noise) [5]. Criteria for the selection of one of these stimulus categories are essentially the options of stimulus manipulation (e.g., real objects will hardly allow for conflicting stimuli) and the relation of internal and external validity. The advancement of virtual reality provided experimenters with extended and promising options for manipulating complex, naturalistic stimuli. Since the virtualization of real environments is known to affect various perceptual and cognitive features [20–23], the impact of virtualization has become another prominent research issue.

The perception of distance and room size in the extrapersonal space depends on particular auditory and visual cues provided by the specific scene. Acoustic distance cues are weighted variably and comprise the sound pressure level and the direct-to-reverberant energy ratio [24–26], spectral attenuation due to air absorption [27], spectral properties due to temporal and directional patterns of reflections of surrounding surfaces [25], as well as spectral alterations due to both near-field conditions and the listener's head and torso. Interaural level and time differences also appear to play a role, namely in connection with orientations and motions of the sound source and the listener [28–30].

In real acoustic environments, perceived egocentric distances are known to be compressed above distances of 2 to 7 m [27, 28, 31–33], hence they are found to be compressed comparably or even more in virtual acoustic environments [32, 34–37]. However, a largely accurate estimation in high-absorbent and an overestimation in low-absorbent virtual environments were also reported [18, 38].

Acoustic room size cues comprise the room-acoustic parameters clarity (C80, C50) [39–41], definition (D50) [41], reverberation time (RT) [39, 42, 43], and likely the characteristics of early reflections [39]. In the medium- and large-sized rooms, the perceived room size was shown to be decreased by a binaural reproduction of the acoustic scene compared to listening in situ [40]. A more recent study found, however, that auralization by dynamic binaural synthesis did not affect the estimation of room size [38].

The estimation of the egocentric distance and the dimensions of visual rooms is based on visual depth cues. Common classifications differentiate between pictorial and non-pictorial, monocular and binocular, as well as visual and oculomotor cues. The cues cover different effective ranges: the personal space (0–2 m), the action space (2–30 m) and/or the vista space (> 30 m) [44]. The non-pictorial depth cues comprise three oculomotor cues: *Convergence* refers to the angle between the eyes' orientation towards the object, *accommodation* to the adaptation of the eye lens' focal length, and *myosis* to the pupillary constriction. *Convergence* is the only binocular oculomotor cue. *Myosis* is effective only within the personal space. Further important non-pictorial visual depth cues are *binocular parallax* (also termed binocular/retinal disparity) referring to differences between the two retinal images due to the permanently different eyes' positions, and *monocular motion (movement) parallax* referring to subsequently different retinal images due to head movements. These cues are effective in both the personal and the action space. Pictorial depth cues are always monocular and based on the extraction of features from the specific images and, where applicable, experiential knowledge. *Linear perspective, texture gradient, overlapping (occlusion), shadowing/shading, retinal image size, aerial perspective* and the *height in the visual field* appertain to the most important pictorial depth cues (see [44–46] for an overview).

In real visual environments, distances are normally estimated much more precisely and accurately than in real acoustic environments [47]. Beyond about 3 m distances are increasingly underestimated both under reduced-cue conditions [48] and in virtual visual environments, no matter if head-mounted displays or large screen immersive displays are used [38, 49–55]. However, also largely accurate estimates in virtual visual environments were reported [18, 56]. While the parallax and the observer-to-screen distance [57], as well as stereoscopy, shadows, and reflections [58] were identified to influence the accuracy of distance estimates in virtual visual environments, the restriction of the field of view [59] and the focal length of the camera lens [60] did not take effect. Room size was observed to be overestimated more in a real visual environment than in the correspondent virtual environment [38], as well as underestimated in other virtual visual environments [18].

Turning to acoustic-visual conditions, the experimental combination of acoustic and visual stimuli can be either congruent or divergent regarding positions or other properties. The widely-used variation of the *presence* of congruent stimulus components (acoustic/visual/acoustic-visual) may be referred to as a co-presence paradigm. A divergent combination independently varies the acoustic and visual *properties* of an acoustic-visual stimulus and is commonly referred to as a conflicting stimulus paradigm.

Under congruent conditions, as experienced in real life, distance estimation is normally highly accurate. Using virtual sound sources and photographs, the additional availability of visual distance information was demonstrated to improve the linearity of the relationship between the physical and the perceptual distance, and to reduce both the within- and the between-subjects variance of the distance judgments [61]. However, virtual acoustic-visual environments may, like virtual visual environments, be subject to compressed distance perception [32], regardless of the application of verbal estimation or perceptually directed action as a measurement protocol [36, 37]. A perceptual comparison between mixed and virtual reality [62] showed that the virtualization of the visual environment increased "aurally perceived" distance and room size estimates (p. 4). The perceived room width was found to be underestimated under the visual, overestimated under the acoustic, and well-estimated under the acoustic-visual conditions [17]. Findings on the accuracy of room size perception are in the same way inconsistent for acoustic-visual environments, as they are for visual environments (see above) [18, 38].

Experiments applying the conflicting stimulus paradigm are normally both more challenging and more instructive [36]. Such experiments have revealed that the localization of an auditory-visual object is largely determined by its visual position, which becomes particularly obvious when compared to the localization of an auditory object. This phenomenon was investigated relatively early [5], and in the case of a lateral or directional offset in the horizontal plane, it was initially referred to as the *ventriloquism effect* ([6], pp. 360-2, [63, 64]). This term has been used in a more abstract sense since, refers to both the spatial and the temporal domain, as well as both directional and distance offsets. The respective effects and aftereffects have been extensively studied (see [65] for an overview).

In the case of an egocentric distance offset, the phenomenon was initially termed the *proximity image effect*: In 1968, Gardner reported that in an anechoic room, the perceived distance was fully determined by the distance of the only visible nearby loudspeaker [7]. A modified replication showed that the effect occurred also when the acoustic distance was nearer than the visual distance, and was only slightly weakened by the chosen semi-reverberant conditions [66]. Zahorik did, however, not observe a clear *proximity image effect* in his replication [67]. Rather, auditory-visual perception, allowing also for prior inspection of the potential sound source locations, improved judgment accuracy when compared to auditory perception (see also [33]). The lack of support for a strict visual dominance in auditory-visual distance localization suggested that sensory modalities contribute to localization with scalable weights.

Indeed, it has been demonstrated that both visual and acoustic stimulus displacements cause significant changes in egocentric distance estimates [68], indicating that visual and auditory influences occur at the same time, however, with different weights. Regarding auditory features, Postma and Katz varied both visual viewpoints and auralizations in a virtual theater, while asking experienced participants for ratings upon distance and room acoustic attributes [69]. Few attributes (including auditory distance) were significantly influenced by the visual contrasts, whereas most attributes were by the acoustic. Interestingly, a deeper data analysis allowed partitioning participants into three groups being mainly susceptible to auditory distance, loudness, and none of the features, respectively, when exposed to different visual conditions. Amongst others, the study points to the principle, that acoustic and visual information weigh normally highest on auditory and visual features, respectively.

In the course of the advancement of a probabilistic view, it was evidenced that the weights adapt to the reliabilities of the sensory estimates in a statistically optimal manner [70]. Maximum Likelihood Estimation (MLE) modeling was shown to apply to different multisensory localization tasks [47, 71–73]. Therefore, acoustic-visual stimuli should generally yield a more precise localization than merely acoustic or visual stimuli [72]. The weights may either be experimentally reduced by adding noise to the stimuli, or in turn, if estimated otherwise, indicate the relative acuity of the stimuli and the reliability of their sensory estimates, respectively. For instance, due to missing or largely reduced interaural level difference and interaural time difference cues, auditory positional information has a lower weight in case of a directional or depth offset in the median plane; in this case, localization is therefore more prone to the influence of visual positional information than in the case of a lateral offset [9, 74]. It was found that acoustic and visual contributions are not symmetric about frontal distance: Using LEDs and noise bursts, a "visual capture" effect and a respective aftereffect in frontal distance perception was observed, with a relatively greater visual bias for visual stimulus components being closer than the acoustic components ([75], p. 4).

Combining MLE with Bayesian causal inference modeling [76] is based on the idea that increasing temporal or spatial divergences between sensory-specific stimuli make the perceiver's inference of more than one physical event more likely, and that multisensory integration takes place only for stimuli subjectively caused by the same physical event. A recent study demonstrated, however, a higher weight of visual signals in auditory-visual integration of spatial signals than predicted by MLE, which might be due to the participants' uncertainty about a single physical cause [77]. While the result of the causal inference is normally not directly observable, the perceived spatial congruency is: Using stereoscopic projection and wave field synthesis, André and colleagues presented participants with 3D stimuli (a speaking virtual character) containing acoustic-visual angular errors. As expected, a higher level of ambient noise (SNR = 4 dB A) caused a 1.1° shift of the point of subjective equivalence and a steeper slope (-0.077 instead of -0.062 per degree) of the psychometric function. Results were not statistically significant, arguably due to the still too high SNR [78].

Evaluating different variants of probabilistic models through experiments using a virtual acoustic-visual environment and applying a dual-report paradigm, the Bayesian causal inference model with a probability matching strategy was found to explain the auditory-visual perception of distance best [79]. The authors also calculated the sensory weights for visual and auditory distances and found that in windows around the correspondent physical distance, auditory distances were predominantly influenced by visual, while visual distances were slightly influenced by auditory sensory estimates. Visual-auditory weights ranged from 0 to 1, auditory-visual weights from 0 to 0.2. Another study showed a major influence of the acoustic properties of spatial scenes on the collective egocentric distance perception (probably due to a substantially restricted visual rendering), whereas room size perception predominantly relied on the visual properties. The virtual environment was based on the dynamic binaural synthesis, speech and music signals, stereoscopic still photographs of a dodecahedron loudspeaker in four rooms, and a 61″ stereoscopic full HD monitor with shutter glasses [18].

The cited studies applied different data collection methods (e.g., triangulated blind walking, absolute scales, 2AFC), virtualization concepts (no virtualization, direct rendering, numerical modeling), stimulus content types (e.g., speech, noise; LEDs, visible sound sources), visual moves (photographs, videos), stimulus dimensionalities (2D, 3D), and reproduction formats (e.g., monophonic sound, sound field synthesis; head-mounted displays, large immersive screens). Thus, connecting the results in a systematic manner is challenging. Findings on the influences of concrete physical properties on percepts and their parameters have not achieved consistency.

Following a research strategy from the general to the specific, the present study focuses on the influences of the acoustic and visual environments' properties in their totality. To this end, whole rooms and source-receiver configurations were experimentally varied. To make this feasible, a collective instead of an individual testing approach was taken, i.e., identical test conditions were allocated not to different repetitions (as necessary for data collection in the context of probabilistic modeling) but to different participants. To emphasize external validity and step towards "naturalistic environments" ([65], p. 805), two prototypic types of content (music, speech), six physically existing rooms, direct 3D renderings, long and meaningful stimuli, and a perceptually validated virtual environment were applied.

1.3 Research questions and hypotheses

Methodologically, the prominent co-presence paradigm entails two restrictions. Firstly, the comparison between the acoustic or visual and the acoustic-visual condition involves two sources of variation: (a) the change between the stimulus' domains (acoustic vs. visual), and (b) the change between the numbers of stimulus domains (1 vs. 2)—i.e., between two basic modes of perceptual processing. Thus, the co-presence paradigm confounds two factors at the cost of internal validity. Since single-domain (acoustic, visual) stimuli do not require a multimodal trade-off, whereas multi-domain (acoustic-visual) stimuli do, different weights of auditory and visual information depending on the basic mode of perceptual processing are expected [79]. To take account of the sources of variation, two dissociating research questions (RQs) were posed.

As a second restriction, the co-presence paradigm does not cover variations within the multi-domain stimulus mode, though it is prevalent in everyday life. Hence, additional RQs ask for the effects of the *properties* of acoustic and visual environments. The respective hypotheses were tested based on six performance rooms with particular source-receiver arrangements, and of both music and speech performances.

RQ 1: To what extent do the perceptual estimates depend on the stimulus domain (acoustic vs. visual, and thereby of the involved modality) as such?
$H1_0$: $\mu_A = \mu_V$.

RQ 2: To what extent do the perceptual estimates depend on the basic mode of perceptual processing (single vs. multi-domain stimuli)?
$H2_0$: $2 \cdot \mu_{AV} = \mu_A + \mu_V$.

RQ 3: To what extent do the perceptual estimates depend on the complex acoustic properties of the multi-domain stimuli?
$H3_0$: $\mu_{A1V\bullet} = \mu_{A2V\bullet} = \mu_{A3V\bullet} = \mu_{A4V\bullet} = \mu_{A5V\bullet} = \mu_{A6V\bullet}$.

RQ 4: To what extent do the perceptual estimates depend on the complex visual properties of the multi-domain stimuli?
$H4_0$: $\mu_{A\bullet V1} = \mu_{A\bullet V2} = \mu_{A\bullet V3} = \mu_{A\bullet V4} = \mu_{A\bullet V5} = \mu_{A\bullet V6}$.

RQ 5: To what extent do the perceptual estimates depend on the interaction of the complex acoustic and visual properties of the multi-domain stimuli?
$H5_0$: $\mu_{AjVk} = \mu_{AjV\bullet} + \mu_{A\bullet Vk} - \mu_{A\bullet V\bullet}$ with $1 \leq j \leq 6$ and $1 \leq k \leq 6$.

Note that not only distance and room size cues but whole scenes were varied, to infer the effects of the entire physical properties of the performance rooms, and therefore of the sensory modalities as such in the context of these environments. RQs 3–5 were made comparative by asking to which extent acoustic and visual properties, and their interaction, do proportionally account for the estimates. For this purpose, commensurable ranges of the factors had to be ensured (2.3, 2.7).

Dependent variables were the perceived egocentric distance and the perceived room size. Where reasonable, the accuracy of the estimates about the physical distances and sizes was also considered.

2. Method

2.1 Methodological considerations and terminology

Answering RQs 1 to 2 requires the application of the co-presence design paradigm. Auralized, visualized, and auralized-visualized spatial scenes are levels of one factor.

Answering RQs 3 to 5 requires the acoustic and visual properties of the scenes to be independent factors rather than just levels of one factor, i.e., the application of the conflicting stimulus paradigm. To allow for the quantification of the proportional influences of acoustic properties, visual properties, and their interaction on the perceptual features, however, certain methodological criteria have to be met, because light and sound cannot be directly compared due to their different physical nature. In particular, not only spatiotemporal congruency but also "crossmodal correspondences" (involving low-level features) ([80], p. 973) as well as semantic congruency [80] of the acoustic and visual stimuli being based on the same scenes (which therefore 'sound as they look'), and the qualitative and quantitative commensurability of the acoustic and visual factors are all required. To this end, the single-domain (acoustic and visual) stimuli have to be derived from the same set of multi-domain (acoustic-visual) stimuli and must be varied in their entirety, i.e., categorically [81].

These considerations result in the need for preservation of all perceptually relevant physical cues and a direct rendering, which we distinguish from fully numerical or partly numerical (hybrid) simulations. The latter approaches are based on assumptions of the physical validity of parametrized material and geometrical room properties, the imperceptibility of structural resolution limits, and/or the physical validity of the applied models on sound and light propagation, including methods of interpolation. By using the term direct rendering, we indicate that the rendering data corresponding to all supported participants' movements were acquired in situ, i.e., neither calculated from a numerical 3D model nor spatially interpolated (see 2.5.).

With the objective of a clear description of investigated effects, it is indicated to factually and terminologically differentiate between ontological realms (*physical*, *perceptual*), and therein between both physical domains (*acoustic*, *visual*; elsewhere also termed *acoustic* and *optical*) and perceptual modalities (*auditory*, *visual*), as well as between modal specificities (*unimodal*, *supramodal*; also referred to as *modal* and *amodal* [80]) [81].

2.2 Perceptual features

In view of both the context of the study (artistic renditions, performance rooms) and the complex variation of the stimuli (2.1), the collection of values of various features was of interest. Accordingly, a differential was used. A superordinate objective of the research project is a comparison of the features regarding their respective dependencies on the presences and properties of the acoustic and visual stimuli. Hence, the questionnaire consisted of 21 perceptual features, subdivided into four sets: auditory features (e.g., *reverberance*), visual features (e.g., *brightness*), aesthetic and presence-related auditory-visual features (e.g., *pleasantness, spatial presence*), and geometric auditory-visual features (*source distance, source width, room length, room width, room height*). Following [82, 83], reference objects (quartet/speaker, room) of the visual and the geometric features were specified. The features were operationalized by bipolar rating scales which were displayed on a tablet computer. Data were entered using touch-sensitive, graphically continuous sliders with a numerical resolution of 127 steps. The geometric feature scales specified units [m] and ranged from 0 to 5 m (*source width*), to 25 m (*source distance, room height*), to 50 m (*room width*), and to 100 m (*room length*). Interval scaling was assumed. The original test language was German. Both the perceived distance and the perceived room size are supramodal (amodal) features by definition [80, 81]. Since optimal preconditions for crossmodal binding and bisensory integration had been established by ensuring crossmodal correspondences and semantic congruency [80, 84], and since they are constant across the co-presence variation and to a considerable extent constant across the conflicting stimulus variation, auditory-

Measure	Perceived length $\hat{L}$	Perceived width $\hat{W}$	Perceived height $\hat{H}$	Perceived source distance $\hat{D}$
Mean	45.833	22.162	14.672	10.431
Standard error of mean	0.905	0.542	0.229	0.185
Cronbach's Alpha	0.926	0.889	0.867	0.850

Table 1.
Comparison of descriptives and internal consistencies of the unidimensional perceptual features. Calculations are based on the total sample (music and speech group, N = 88) and all rooms under the mere visual conditions (V1–V6). The conditions were pooled for the calculation of mean and standard error, and treated as separate items for the calculation of Cronbach's alpha.

visual integration was assumed to be able to occur either automatically or intentionally. Hence, participants were asked to estimate values of unitary features. No problems concerning this task were reported. Because test participants do not maintain linearity when assessing three-dimensional room volume using a single one-dimensional scale [18], they were asked for separate length ($\hat{L}$), width ($\hat{W}$), and height ($\hat{H}$) estimates.

Since the visual stimuli showed only a part of the frontal hemisphere (see 2.5), the participants had to base their assessment of the invisible rear part of the rooms' length on the visible frontal length, the room shape, their position in the room, and their experiential knowledge on the shape and size of performance rooms. Hence, before analyzing the calculated room volume/size estimates, dispersion and reliability measures of the unidimensional perceptual features were inspected (**Table 1**).

Neither the reliability nor the dispersion of the perceived length is conspicuous, since the values for Cronbach's Alpha are throughout high, for the perceived length even excellent, and the error-to-mean ratios are consistent across the perceptual features. By calculating the cube root of the product of the three collected features, the one-dimensional feature *perceived room size* $\hat{S}$ was derived. This report focuses on the *perceived source distance* ($\hat{D}$) and the *perceived room size* ($\hat{S}$).

2.3 Design

Since answering RQs 1 to 2 requires the application of the co-presence paradigm, the factor *Domain* was defined by the levels auralized (**A**), visualized (**V**), and auralized-visualized (**AV**). To raise the external validity of the potential main effects and to allow for the observation of room-specific effects, the second factor *Room* was introduced, comprising six different performance rooms under examination (levels **R1** to **R6**, see **Table 2** for specific labels). Answering RQs 3 to 5 requires the application of the conflicting stimulus paradigm. Thus, the factor *Auralized room* was defined by the acoustic stimulus components of the six rooms (levels **A1** to **A6**), and the factor *Visualized room* by the respective visual stimulus components (levels **V1** to **V6**). An integrative survey design covered both the co-presence and the conflicting stimulus paradigms while avoiding a redundant presentation of **AV** congruent stimuli across the paradigms. To limit the total sample to a practicable size, these four factors had to be realized as within-subjects factors. **A** and **V** stimuli were presented first, followed by **A**i-**V**j (including **AV**, i.e., $i = j$) stimuli. Within these two test partitions, the stimuli were presented in individually randomized order. By introducing the between-subjects factor *Content*, the total sample was divided into two groups assigned to the music and speech renditions, respectively.

The number of trials within a test sequence corresponds to the number of experimental conditions (factor level combinations). There are two options for

Label	KH	RT	KO	JC	KE	GH
Name	Konzert-haus	Renais-sance Theater	Komi-sche Oper	Jesus-Christus-Kirche	Kloster Eberbach	Gewandhaus
Function	Concert hall	Theatre	Opera	Church	Church	Concert hall
Volume V [m³]	1899	1903	7266	8079	19539	22202
Size S [m]	12.383	12.392	19.369	20.066	26.934	28.106
Position of receiver (row no./seat no.)	6/8–9	11/178	9/20	3/-	-/-	6/9
Distance receiver— central source D [m]	9.97	9.90	9.46	7.19	15.84	9.84
Absorption coefficient $\alpha_{mean(Sabine)}$	0.18	0.20	0.30	0.17	0.02	0.28
Reverberation time $RT30_{mid}$ [s]	1.29	0.80	1.31	2.81	7.92	2.29
Early Decay Time EDT_{mid} [s]	1.31	0.72	1.17	2.67	8.20	1.99

Table 2.
Geometric and material properties of the selected rooms (taken from [85]). The index mid refers to the mean of two-octave bands (500 Hz, 1 kHz).

allocating the trials to the scale items: (a) A long stimulus (ca. 2:00 min, cf. 2.5) is judged by means of the 21 items (2.2); there is just one test sequence. (b) A short stimulus (ca. 6 sec) is judged by means of one feature; the number of test sequences corresponds to the number of features. Option (a) was chosen for the following reasons: (1) In the case of option (b), the comparison of the features, as required by the research project (2.2), would be confounded with the repetition of a stimulus, including greater time intervals, whereas it is not in case of option (a). (2) Short stimuli would run counter to both the context (1.1) and the methodological aim (2.2, 2.3) of the study: artistic renditions are much longer than a few seconds, and— particularly regarding the aesthetic and presence features—responses to very short extracts could not be generalized for entire renditions. (3) To yield valid responses, stimuli must provide enough time and information for judgment formation. Build- ing up an aesthetic impression about very short extracts of an artistic rendition would be hardly possible due to the lack of information about the course of time. Thus, artistically self-contained sections were to be presented at least. Long stimuli provide a greater number and variety of physical events, so that each participant can rely on the individually most helpful cues. (4) In the case of option (a) the decision times vary and are unknown, i.e., within the samples, decision times, as well as causal events and their cues, are pooled. On the one hand, this increases the external validity. On the other hand, it also decreases the internal validity, though, to an acceptable level, since both physical distance and size are constant within each stimulus, and attribution of the estimates to detailed cues or events is not part of the research questions (cf. 1.3).

2.4 Sample

The required sample size was calculated a priori with the aid of the software package G*POWER 3 [85, 86]. Since the groups of the factor *Content* were analyzed separately, only full-factorial repeated measures designs were considered. The sample size had to be geared to the small 3×6 co-presence design. To statistically reveal a relatively small effect size ($f = 0.15$) at a type I error level of $\alpha = 0.05$ and a test power of $1-\beta = 0.95$ while assuming a correlation amongst the repeated measurements of $r = 0.6$ and an optional nonsphericity correction of $\varepsilon = 0.7$, the minimum sample size per group accounted for $n = 38$. A total of 114 subjects being affine to music per self-report were initially recruited for the experiment. Subjects were excluded in the following cases (multiple incidences possible):

- Hypoacusis; criterion: audiogram, hearing threshold >20 dB HL at either ear at any of seven tested frequency bands (125 to 8000 Hz), uncompensated by hearing aid (0 subjects).

- Vision deficits; criterion: self-reported deficits, uncompensated by visual aid (0 subjects).

- Red and/or green color blindness; criterion: unpassed Ishihara tests for protanomaly and deuteranomaly (3 subjects).

- Loss of stereopsis; criterion: unpassed contour stereopsis test using the shutter glasses of the projection system (4 subjects).

- Technical incident; failure of saving response data (6 subjects).

- Subjectively untrue responses; criterion: implausible perceptual bias (factor ≥ 5) with reference to visual geometric dimensions (14 subjects, most frequent response: "0 m").

The resultant valid net sample sizes accounted for $n = 50$ for the music group and for $n = 38$ for the speech group, comprising 32 female and 56 male voluntary non-experts aged from 21 to 65 years. The frequencies of the participants within the age classes (20s, 30s, 40s, 50s, 60s) amount to $f_{abs} = \{36; 24; 13; 10; 5\}$. Participants did not receive incentives.

2.5 Stimuli

As far as possible in a virtual environment, a maximum ecological validity of the stimuli was sought by selecting dedicated performance rooms, artistic content and professional music and speech performers.

Six performance rooms differing in volume (low, medium, high) and average acoustic absorption coefficient (low: $\alpha_{mean(Sabine)} < 0.2$; high: $\alpha_{mean(Sabine)} \geq 0.2$) were selected. Taking into account good speech intelligibility and an accurate perceptibility of the physical room properties (e.g., the visibility of the ceiling height), optimum receiver positions were defined. Based on geometric measures acquired in situ, models of the interior spaces, including the source-receiver-arrangements, were built using the software *SketchUp* (by Google/Trimble) and the plugin *Volume Calculator* (by TGI). The volumes and surface areas of the rooms were then calculated. Standard acoustic measures were taken in situ, in dependence on DIN EN ISO 3382-1 [87]. To corroborate the rooms' selection according to the absorption criterion ex post, Sabine absorption coefficients were calculated from the reverberation times and the geometric properties [88]. The air absorption effect was included; attenuation coefficients were taken from [89]. **Table 2** presents geometric and material properties. Distances were measured directly (i.e., not necessarily in the horizontal plane) from the acoustic center of the central sound source to the interaural center of the head and torso simulator; they all cover the extrapersonal space. Detailed acoustic measurement reports (research data) are available [90].

The artistic content comprised a musical work and a text, which were chosen to support the perceptibility of the specific room properties by featuring, e.g., impulsivity and sufficient pauses. Two-minute excerpts of Claude Debussy's String Quartet in g minor, op. 10, 2nd movement, and of Rainer Maria Rilke's 1st Duino Elegy were selected. The artistic renditions were audio recorded in the anechoic room of the Technische Universität Berlin.

The performances were presented in the Virtual Concert Hall at Technische Universität Berlin, providing virtual acoustic and visual 3D renditions in rooms. It was particularly designed to meet the methodological requirements (2.1, 2.3), and was completely based on directional binaural room impulse responses (BRIRs) and stereoscopic panoramic images acquired in situ by means of the head and torso simulator *FABIAN* [91, 92]. The stimulus reproduction applied dynamic binaural synthesis by means of an extraaural headset and a semi-panoramic active stereoscopic video projection featuring an effective physical resolution of 4812×1800 pixels (**Figure 1**).

The used BRIRs contained the fixed HRTFs of *FABIAN*, hence non-individual HRTFs with regard to the listeners. Experimentation showed that head tracking in connection with non-individual HRTFs improves externalization [93], virtually eliminates front/back confusion, and substantially reduces elevation errors [94]. The auralization system used for this study included head tracking with an angular resolution of 1° and an angular range of ±80° which had to be proved sufficient [95, 96]. It also compensated for spectral coloration [97]. Experimentation also showed that non-individual HpTF compensation, as applied for the present study, outperforms individual HpTF compensation in the specific case of non-individual binaural recordings [98]. System latency was minimized to a level below the perceptual threshold [99]. Cross-fade artifacts were reduced by the applied rendering

Figure 1.
Participant in the Virtual Concert Hall (visual condition: KO).

algorithm fwonder. The system also allowed for the adaption to the participants' individual ITDs [100].

The virtual environment did not provide auditory motion parallax cues by supporting lateral motion interactivity and rendering. This was due to limited in-situ acquisition times in the performance rooms. It would have required measurements at several additional positions of the head and torso simulator, depending on the content-specific minimum audible BRIR grid [101, 102], and thus would have multiplied the expenditure of acquisition time beyond the rooms' availability. However, auditory motion parallax, describing the change in the angular direction of a distant sound source due to the movement of the listener, is assumed to be a supporting cue in absolute distance estimation [103] and known to be a cue in relative depth estimation [104]. Regarding a distance range within the personal space, it was demonstrated by means of a depth discrimination task, and under exclusion of all other distance cues, that auditory motion parallax is exploited by listeners allowing for the perception of distance differences of unknown acoustic stimuli [104]. The cue was shown to be effective for distances between 0.3 and 1.0 m and to be exploitable for lateral head movements within a range of 46 cm. The participants' sensitivity was highest during self-induced motion. Even sensitive subjects did not perceive distance differences corresponding to angular displacements below 3.2°. This value is higher than the minimum audible movement angles (MAMAs) found in previous research (see [105] for an overview). Regarding a distance range of 1 to 10 m, Rumukkainen and colleagues determined the self-translation minimum audible angle (ST-MAA) to be 3.3° by means of 2AFC discrimination tasks without an external reference [106]. Taking into account the absence of external references in the present study and applying the ST-MAA to the nearest sound source used (7.19 m), a concertgoer would remain below perceptual threshold within a lateral moving range of ±41.5 cm, which corresponds to 150% of a typical concert seat's width. Respective lateral movements are normally not observed amongst visitors of classical concerts. Since a relative lateral shift of the listener above the perceptual threshold is a precondition for yielding distance information from the auditory motion parallax cue by triangulation, we do expect neither an appreciable bias nor a deterioration of the accuracy of distance perception introduced by the absence of lateral motion interactivity and rendering.

As a result, the Virtual Concert Hall at Technische Universität Berlin provided almost all relevant auditory cues without major biases (rich-cue condition). Exceptions are the missing supports for (rarely performed and normally small) head orientations around the pitch and roll axes.

The sound pressure level of the virtual rendition was adjusted to the sound pressure level of a live rendition of a string quartet in a real room, which was recorded by the calibrated head and torso simulator. Accounting for the gain of the signal chain and the rooms' STI measures, the level of the scenes' average sound pressure level at the blocked ear canal was L_p = 72.5 dB SPL for a selected *mezzoforte* passage. Likewise, the speech's sound pressure level was adapted to a rendition in a real room and averaged out at L_p = 59.5 dB SPL for a moderate declamatory dynamics stage.

The acquisition of the visual rendering data applied a fixed stereo base, which does not necessarily accord with the participants' individual interpupillary distances (IPDs). Respective differences might potentially bias the individual distance and room size perception. To date, experimentation has shown inconsistent effects of the variation of IPD differences on distance perception (see [46] for a review). Most studies cannot be translated into the present study, since they investigated maximum target distances of 1 m and/or used simple numerically modeled objects/environments. Moreover, results differ regarding the significance, the size and/or the direction of the effects. This is apparently due to different rendering technologies (stereoscopic projection, HMD, CAVE), stages of virtualization (mixed reality, virtual reality), target distances (personal space, action space), simulated objects/environments (simple graphic objects, shapes, persons in hallways), and measurement protocols (triangulated distance estimation, blind walking, visual alignment, verbal estimation) [107–113]. Few experiments investigated distances roughly similar to those used in the present study (about 7 to 16 m). While Willemsen and colleagues did not observe a significant effect of IPD individualization on distance judgments [114], a large variation of the stereo base (0 to 4 times the IPD) showed significant effects on both distance and size judgments: Greater stereo bases resulted in perceptually closer and smaller objects [115]. However, relevance for the descriptive measures, effect sizes and significances of the present study is given rather by the expected value and distribution of the IPD differences than by their individual values. Anthropometric data of the German resident population, from which the sample was drawn, state median IPDs of 61 mm (male persons) and 60 mm (female persons) within the age range of 18 to 65 years [116]. Since the values do nearly exactly meet the stereo base of the target acquisition (60 mm), a substantial collective perceptual bias is unlikely to occur.

Limitations of the visual rendering pertain to the field of view (161° × 56°), which should at least not affect distance perception [59, 117]; the angular resolution (2.1 arcmin), which might affect distance perception [57]; the fixed single focal plane in stereoscopy providing an invariant accommodation cue, so that the connection between convergence and accommodation is suspended [45]; and an undersized luminance of the projection. Data projectors could not provide the luminance and the contrast of the real scenes, especially in connection with shutter glasses. Thus, the luminances of the scenes were fitted into the projectors' dynamic range while maintaining compressed relations of the luminances. Scene luminances were calculated from exposure time, aperture, and ISO arithmetic film speed of correctly exposed photographs of a centrally placed and vertically oriented 18% gray card according to the additive system of photographic exposure (APEX). The average loss of the luminance value Bv introduced by the projection and shutter glasses was 2.88. The average scene luminance L_v of the gray cards amounted to 0.82 cd/m^2. Detailed information regarding room acquisition, content production, and stimulus reproduction for the Virtual Concert Hall was published separately [118].

Since electronic media transform both the physical stimuli and their perception, the replacement of natural by mediatized stimuli for serious experimental purposes demands the knowledge of the perceptual influences of the applied mediatizing system, as also pointed out by [16, 21]. The rendering technique of the Virtual Concert Hall was shown to provide perceptually plausible auralizations [119]. Specifically, the Virtual Concert Hall at Technische Universität Berlin was subjected to a test of auditory-visual validation by comparing a real scene and the correspondent virtual scene [38]. Amongst others, it yielded nearly equal loudness judgments of the real and the virtual environment, whereas the virtual environment—apparently due to the dark surrounding—was perceived slightly brighter than the respective real environment. The virtualization also generally lowered the perceived source distance and the perceived size of a real room—mainly due to the visual rendering. The mere auditory underestimation of source distance and room size introduced by the virtualization amounted only to 6.6 and 1.9%, respectively. The biases are considered in the discussion section.

2.6 Procedure

Each participant ran through the test procedure individually. The procedure lasted about 3 hours and 10 minutes, and comprised color vision and stereopsis tests, audiometry, a socio-demographic questionnaire, a privacy agreement, the clarification of the questionnaire, the measurement of the individual inter-tragus distance (necessary for the technical adaption to the individuals' ITDs), cabling, a familiarization sequence, and the actual test runs, inclusive of self-imposed breaks.

2.7 Data analysis

Arithmetic means standard deviations (**Tables 11** and **12**) and standard errors were calculated for all combinations of factor levels. The means were plotted against the combinations. According to the test design (2.3), the co-presence paradigm required 3×6 repeated measures analyses of variance (rmANOVA), the conflicting stimulus paradigm 6×6 rmANOVA for either level of *Content*. *Content* was not regarded as a factor for analysis because it was not covered by the RQs, and the quantification of the proportions according to RQs 3–5 were to be made possible separately for both music and speech. Kolmogorov-Smirnov tests indicated that the assumption of normally distributed error components was met with the exceptions of source distance under the conditions speech **A0-V5** ($KS\text{-}Z = 1.390, p = 0.042$) and speech **A6-V3** ($KS\text{-}Z = 1.442, p = 0.031$), and of room size under the conditions speech **A0-V5** ($KS\text{-}Z = 1.500, p = 0.022$), speech **A5-V0** ($KS\text{-}Z = 1.759, p = 0.004$), and music **A4-V5** ($KS\text{-}Z = 1.428, p = 0.034$). The minor violations concerning 4.8% of the conditions were deemed tolerable because of the robustness of the rmANOVA. Mauchly's sphericity tests indicated a significant violation of the sphericity assumption in both the 3×6 and the 6×6 analyses, which was compensated for by correcting the degrees of freedom using Greenhouse-Geisser estimates. To answer RQs 1 and 2, an orthogonal set of planned main contrasts (reverse Helmert) was calculated: Simple contrast **V** vs. **A**; combined contrast **VA** vs. {**V**, **A**}. To allow different approaches to effect size comparison, partial eta squared η_P^2, classical eta squared η^2, and generalized eta squared η_G^2 [120, 121] were reported for the omnibus tests. Because of RQs 3–5, and taking advantage of the commensurability of the factors *Auralized room* and *Visualized room* of the conflicting stimulus design, the η^2 effect sizes were particularly reported as indicators for the proportional influence of the acoustic room properties, the visual room properties, and their interaction on the geometric features. To allow their direct comparison in a simplified manner, the net effect sizes (the proportions of the explained variance) given by $\eta_{X(net)}^2 = \eta_X^2/(\eta_A^2 + \eta_V^2 + \eta_{A \times V}^2)$ were also reported. Based on Cohen's f ([122], p. 281), which was calculated from η^2 ([123], p. 7), the effect sizes were classified as small, medium or large.

3. Results

3.1 Perceived source distance

3.1.1 Co-presence paradigm

Source distance showed significant main and interaction effects of *Domain* and *Room* for both music (**Table 3**) and speech (**Table 4**). Effects were large for *Room* and medium size for *Domain* and *Domain* $\times$ *Room*. The mean distance estimates were generally lower for speech than for music, and the range of the mean estimates introduced by the factor *Domain* was lower for the low-absorbent (wet) and higher for the high-absorbent (dry) rooms, even though it was not hypothesized or tested (**Figures 2** and **3**; **Tables 11** and **12**).

S. o. V.	SS	df_{adj}	MS	F	p	η^2	f	η_G^2	η_P^2	1-β
Domain	1521.061	1.782	853.503	36.965	<0.001	0.086	0.306	0.122	0.430	>0.999
Error (Domain)	2016.285	87.325	23.090							
Room	4610.610	3.845	1199.166	137.464	<0.001	0.260	0.593	0.296	0.737	>0.999
Error (Room)	1643.487	188.397	8.724							
Domain × Room	597.939	7.113	84.059	9.593	<0.001	0.034	0.187	0.052	0.164	>0.999
Error (D. × R.)	3054.229	348.552	8.763							

Table 3.
Results of the rmANOVA for perceived source distance $\hat{D}$ (music, co-presence paradigm).

S. o. V.	SS	df_{adj}	MS	F	p	η^2	f	η_G^2	η_P^2	1-β
Domain	655.466	1.683	389.381	23.350	<0.001	0.058	0.248	0.073	0.387	>0.999
Error (Domain)	1038.621	62.284	16.676							
Room	1712.901	3.387	505.745	41.676	<0.001	0.152	0.423	0.171	0.530	>0.999
Error (Room)	1520.729	125.315	12.135							
Domain × Room	639.600	5.709	112.027	10.836	<0.001	0.057	0.245	0.072	0.227	>0.999
Error (D. × R.)	2183.952	211.245	10.338							

Table 4.
Results of the rmANOVA for perceived source distance $\hat{D}$ (speech, co-presence paradigm).

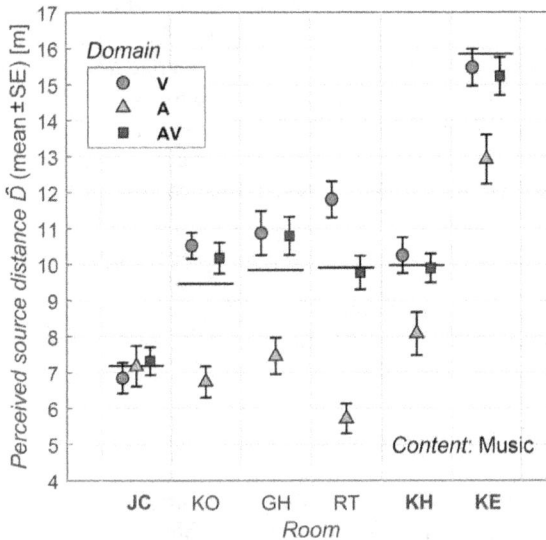

Figure 2.
Means (markers) and standard errors (bars) of perceived source distance $\hat{D}$ against factor levels of Room and Domain for music. Horizontal lines indicate the particular physical source distance D within each room. Bold labels indicate low-absorbent rooms.

Regarding RQ 1, a priori main contrasts indicate that the mean estimates at level V were considerably higher than those at level A. The mean differences account for 2.95 m (music), $F(1,49) = 52.910$, $p < 0.001$, $\eta_P^2 = 0.519$, and for 2.38 m (speech), $F(1,49) = 32.712$, $p < 0.001$, $\eta_P^2 = 0.469$. This is also consistent on a descriptive basis

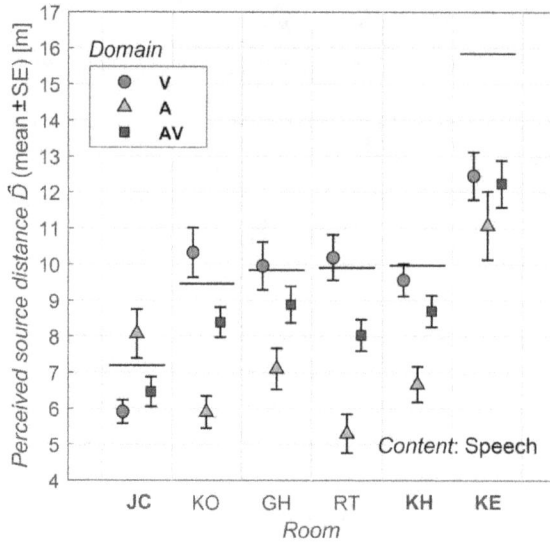

Figure 3.
Means (markers) and standard errors (bars) of perceived source distance $\hat{D}$ against factor levels of Room and Domain for speech. Horizontal lines indicate the particular physical source distance D within each room. Bold labels indicate low-absorbent rooms.

across all rooms except JC, which involves the smallest physical distance ($s = 7.19$ m) and shows a lower mean estimate under the **V** than under the **A** condition for both music and speech. Looking at RQ 2, the mean estimates at level **AV** were higher than the average of the mean estimates at levels **A** and **V**. The mean differences accounted for 1.04 m in the music group (a priori main contrast), $F(1,49) = 13.141$, $p = 0.001$, $\eta_P^2 = 0.211$, and 0.24 m in the speech group (contrast not significant). The **AV** mean estimates were located at 85% of the range between the mean estimates at levels **V** and **A** in the music group and at 60% in the speech group.

Looking at the accuracy of the estimates, the mean estimates differed from the mean physical source distance by -2.36 m (-22.7%) at level **A**, $+0.59$ m ($+5.7\%$) at level **V**, and $+0.16$ m ($+1.5\%$) at level **AV** in the music group, and by -3.02 m (-29.1%) at level **A**, -0.63 m (-6.1%) at level **V**, and -1.59 m (-15.3%) at level **AV** in the speech group. Overall, the physical distances were met best by the estimates at level **AV** in the music group, and by the estimates at level **V** in the speech group.

3.1.2 Conflicting stimulus paradigm

Auralized room and *Visualized room* showed significant main effects on source distance for both music (**Table 5**) and speech (**Table 6**), however, no significant interaction effect. Effects of *Auralized room* were of small size, whereas effects of *Visualized room* were classified as large. Regarding music, $\eta_{A(net)}^2 = 7\%$ of the proportion of the explained variance (see 2.7) arose from *Auralized room*, $\eta_{V(net)}^2 = 91\%$ from *Visualized room*. Under the speech condition, the proportions accounted for 11% (*Auralized room*) and 88% (*Visualized room*).

Figures 4 and **5** show the generally lower mean distance estimates for the speech by trend. The figures also illustrate the ranges of the mean estimates. The average range of mean estimates caused by *Auralized room* was 1.69 m, while the range caused by *Visualized room* accounted for 5.74 m. The range of the physical source

S. o. V.	SS	df_{adj}	MS	F	p	η^2	f	η_G^2	η_P^2	1-β
Auralized room	469.724	5.000	93.945	13.143	<0.001	0.017	0.133	0.023	0.211	>0.999
Error (A. room)	1751.252	131.608	13.307							
Visualized room	6256.608	2.602	2404.324	105.444	<0.001	0.233	0.551	0.238	0.683	>0.999
Error (V. room)	2907.446	127.509	22.802							
A. room × V. room	134.833	13.677	9.858	1.566	0.086	0.005	0.071	0.007	0.031	0.868
Error (A. r. × V. r.)	4219.961	670.192	6.297							

Table 5.
Results of the rmANOVA for perceived source distance $\hat{D}$ (music, conflicting stimulus paradigm).

S. o. V.	SS	df_{adj}	MS	F	p	η^2	f	η_G^2	η_P^2	1-β
Auralized room	375.912	1.667	225.526	9.314	0.001	0.023	0.153	0.028	0.201	0.951
Error (A. room)	1493.259	61.672	24.213							
Visualized room	2936.460	2.724	1077.931	48.375	<0.001	0.178	0.465	0.183	0.567	>0.999
Error (V. room)	2245.993	100.794	22.283							
A. room × V. room	31.620	12.609	2.508	0.531	0.902	0.002	0.044	0.002	0.014	0.317
Error (A. r. × V. r.)	2203.942	466.540	4.724							

Table 6.
Results of the rmANOVA for perceived source distance $\hat{D}$ (speech, conflicting stimulus paradigm).

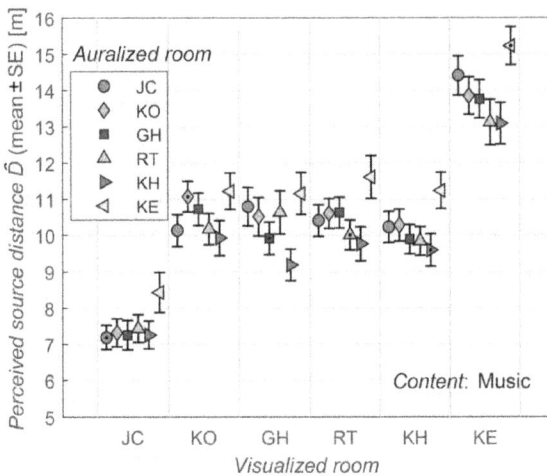

Figure 4.
Means (markers) and standard errors (bars) of perceived source distance $\hat{D}$ against factor levels of Auralized room and Visualized room for music. Dots within markers indicate acoustic-visual congruency.

distance was 8.65 m. As a rule, the auralized room KE led to a maximal mean estimate and the auralized room RT to a minimal mean estimate within each visualized room. In turn, the visualized room KE led to a maximal mean estimate and the visualized room JC to a minimal mean estimate within each auralized room. The mean estimates do not indicate that acoustic-visual congruency as such yielded maximal, minimal or especially accurate mean distance estimates.

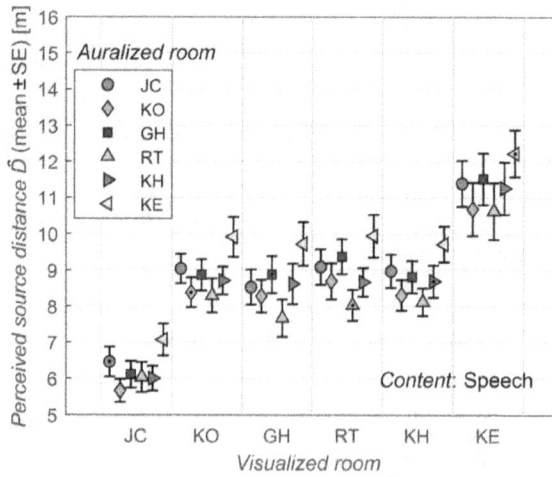

Figure 5.
Means (markers) and standard errors (bars) of perceived source distance $\hat{D}$ against factor levels of Auralized room and Visualized room for speech. Dots within markers indicate acoustic-visual congruency.

3.2 Perceived room size

3.2.1 Co-presence paradigm

Room size showed significant main and interaction effects of *Domain* and *Room* for both music (**Table 7**) and speech (**Table 8**). Effects were of large size for *Domain* (music) and *Room* and of medium size for *Domain* (speech) and *Domain* ×

S. o. V.	SS	df_{adj}	MS	F	p	η^2	f	η_G^2	η_P^2	1-β
Domain	10148.965	1.651	6145.611	70.421	<0.001	0.115	0.361	0.180	0.590	>0.999
Error (Domain)	7061.808	80.919	87.270							
Room	28109.442	3.650	7701.183	226.890	<0.001	0.319	0.685	0.379	0.822	>0.999
Error (Room)	6070.632	178.851	33.942							
Domain × Room	3733.981	7.522	496.424	21.814	<0.001	0.042	0.210	0.075	0.308	>0.999
Error (D. × R.)	8387.358	315.667	26.570							

Table 7.
Results of the rmANOVA for perceived room size $\hat{S}$ (music, co-presence paradigm).

S. o. V.	SS	df_{adj}	MS	F	p	η^2	f	η_G^2	η_P^2	1-β
Domain	6484.837	1.513	4285.200	42.093	<0.001	0.082	0.299	0.131	0.532	>0.999
Error (Domain)	5700.224	55.992	101.803							
Room	26573.103	2.994	8875.557	165.259	<0.001	0.337	0.713	0.383	0.817	>0.999
Error (Room)	5949.496	101.789	58.449							
Domain × Room	3000.770	6.785	442.292	19.791	<0.001	0.038	0.199	0.065	0.348	>0.999
Error (D. × R.)	5610.150	251.030	22.349							

Table 8.
Results of the rmANOVA for perceived room size $\hat{S}$ (speech, co-presence paradigm).

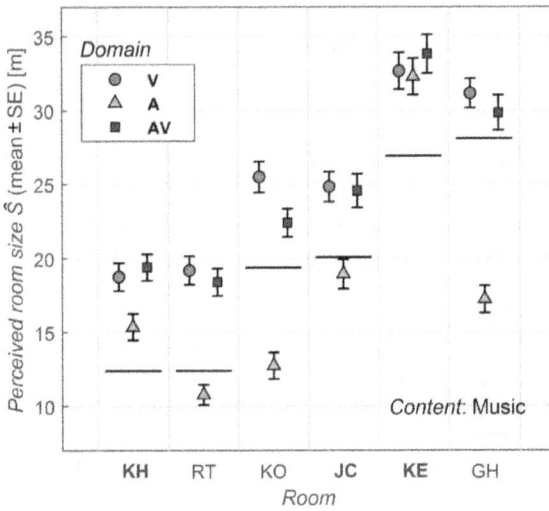

Figure 6.
Means (markers) and standard errors (bars) of perceived room size Ŝ against factor levels of Room and Domain for music. Horizontal lines indicate the particular physical room size S of each room. Bold labels indicate low-absorbent rooms.

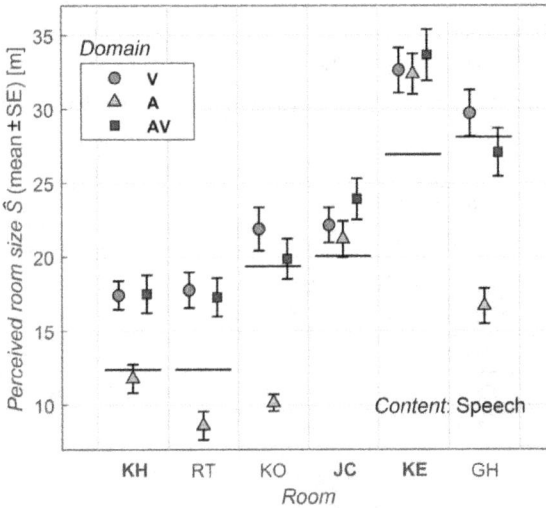

Figure 7.
Means (markers) and standard errors (bars) of perceived room size Ŝ against factor levels of Room and Domain for music. Horizontal lines indicate the particular physical room size S of each room. Bold labels indicate low-absorbent rooms.

Room. The mean size estimates were slightly lower for speech than for music by trend (**Figures 6** and 7).

Regarding RQ 1, a priori contrasts indicated that the mean estimates at level **V** were considerably higher than those at level **A**. The mean differences account for 7.40 m (music), $F(1,49) = 97.748$, $p < 0.001$, $\eta_p^2 = 0.666$, and for 6.71 m (speech), $F(1,49) = 51.457$, $p < 0.001$, $\eta_p^2 = 0.582$. Looking at RQ 2, a priori contrasts showed that the mean estimates at level **AV** were higher than the average of the mean estimates at levels **A** and **V**. The mean differences accounted for 3.11 m (music),

$F(1,49) = 32.124$, $p < 0.001$, $\eta_p^2 = 0.396$, and 2.99 m (speech), $F(1,49) = 24.933$, $p < 0.001$, $\eta_p^2 = 0.403$. The **AV** estimates were located at 92% of the range between the mean estimates at levels **V** and **A** in the music group and at 94% in the speech group.

As with source distance, the range of the mean room size estimates introduced by the factor *Domain* was lower for the low-absorbent (wet) and higher for the high-absorbent (dry) rooms, even though this was not hypothesized or tested.

Accuracies were generally low regardless of the level of *Domain*. The mean room size estimates differed from the mean physical room size by −2.00 m (−10.0%) at level **A**, +5.47 m (+27.5%) at level **V**, and +4.86 m (+24.5%) at level **AV** in the music group, and by −3.08 m (−15.5%) at level **A**, +3.72 m (+18.7%) at level **V**, and +3.34 m (+16.8%) at level **AV** in the speech group. Overall, the physical sizes were generally best approximated by the estimates at level **A**. Specifically, in low-absorbent rooms (KH, JC, KE) and the small dry room (RT), physical room sizes were best approximated by the estimates at level **A**, whereas in medium- and large-sized dry rooms (KO, GH) they were best approximated by the estimates at levels **AV** and **V**.

3.2.2 Conflicting stimulus paradigm

Auralized room and *Visualized room* showed significant main effects on room size for both music (**Table 9**) and speech (**Table 10**), however, no significant interaction effect. Effects of *Auralized room* were of small size, whereas effects of *Visualized room* were classified as large. Regarding music, $\eta^2_{A(net)} = 9\%$ of the proportion of the explained variance (see 2.7) arose from *Auralized room*, $\eta^2_{V(net)} = 90\%$

S. o. V.	SS	df_{adj}	MS	F	p	η^2	f	η^2_G	η^2_p	1-β
Auralized room	3275.179	2.048	1599.570	25.911	<0.001	0.024	0.156	0.031	0.346	>0.999
Error (A. room)	6193.742	100.329	61.734							
Visualized room	32107.238	3.203	10025.415	110.275	<0.001	0.233	0.551	0.239	0.692	>0.999
Error (V. room)	14266.617	156.927	90.913							
A. room × V. room	375.257	12.004	31.262	1.344	0.189	0.003	0.052	0.004	0.027	0.754
Error (A. r. × V. r.)	13678.450	588.172	23.256							

Table 9.
Results of the rmANOVA for perceived room size $\hat{S}$ (music, conflicting stimulus paradigm).

S. o. V.	SS	df_{adj}	MS	F	p	η^2	f	η^2_G	η^2_p	1-β
Auralized room	3799.130	1.446	2626.800	11.517	<0.001	0.026	0.162	0.030	0.237	0.968
Error (A. room)	12205.465	53.513	228.084							
Visualized room	23087.978	2.228	10363.307	54.821	<0.001	0.155	0.429	0.160	0.597	>0.999
Error (V. room)	15582.628	82.431	189.039							
A. room × V. room	185.804	7.540	24.642	0.662	0.716	0.001	0.035	0.002	0.018	0.296
Error (A. r. × V. r.)	10382.097	278.982	37.214							

Table 10.
Results of the rmANOVA for perceived room size $\hat{S}$ (speech, conflicting stimulus paradigm).

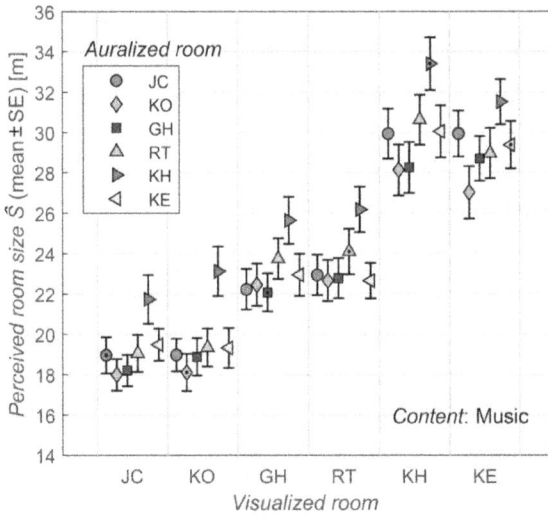

Figure 8.
Means (markers) and standard errors (bars) of perceived room size Ŝ *against factor levels of* Auralized room *and* Visualized room *for music. Dots within markers indicate acoustic-visual congruency.*

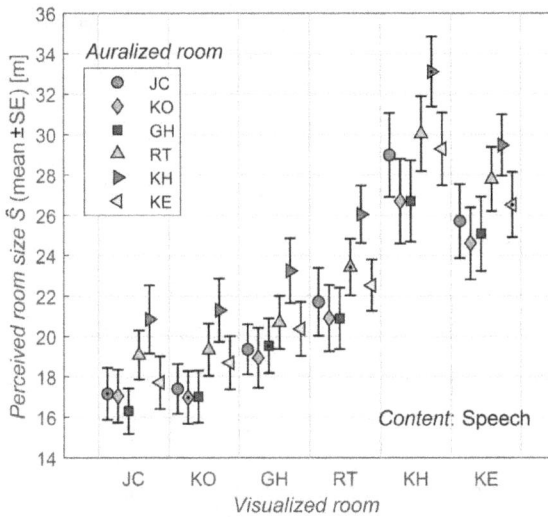

Figure 9.
Means (markers) and standard errors (bars) of perceived room size Ŝ *against factor levels of* Auralized room *and* Visualized room *for speech. Dots within markers indicate acoustic-visual congruency.*

from *Visualized room*. Under the speech condition, the proportions accounted for 14% (*Auralized room*) and 85% (*Visualized room*).

Figures 8 and **9** show the generally lower mean room size estimates for the speech by trend. The figures also illustrate the ranges of the mean estimates. The average range of mean estimates caused by *Auralized room* was 4.54 m, the range caused by *Visualized room* accounted for 10.99 m. The range of the physical room size was 15.72 m. As a rule, the auralized room KE led to a maximal mean estimate and the auralized room RT to a minimal mean estimate within each visualized room. In turn, the visualized room KE led to a maximal mean estimate and the visualized room KH mostly to a minimal mean estimate within each auralized room. The mean estimates do not indicate that acoustic-visual congruency as such yielded maximal, minimal or especially accurate mean size estimates.

4. Discussion

4.1 Presence of auralized and visualized rooms

Most of the results apply likewise to both egocentric distance and room size estimation. RQ 1 asked for the difference between the modalities as such. Mean estimates across the rooms based only on visual information significantly and considerably exceeded those based only on acoustic information, specifically by about a fourth of the mean physical property in the case of distance and by about a third in the case of size. Hence, H1$_1$ can be accepted and might be reformulated directionally (H1$_1$: $\mu_A < \mu_V$) for future experimentation. Regarding egocentric distance estimation, the finding is plausible in principle given the reported compression of distance perception in real acoustic environments [27, 28, 31–33] and virtual acoustic environments [32, 34–36]. However, it does not agree with [36], who observed a compressed perception of visual distances between 1.5 and 5.0 m, or with [18], who used nearly the same auralization system in connection with smaller distances (1.93–5.88 m) and a restricted visualization. Though the general finding $\hat{D}_V > \hat{D}_A$ does also not accord with the finding of [38] under the virtual environment condition ($\hat{D}_V < \hat{D}_A$), the exceptional observation at the smallest physical distance (D = 7.19 m, room JC) does. This is likely to be due to the same physical distance used in [38], indicating that the general finding might be confined to physical distances greater than about 8 m. However, checking the acceptance of inference from virtual rooms to real rooms for the music content by multiplying the mean estimates of the present study by the reality-to-virtuality ratios (*RVRs*) of the mean estimates of [38] ($RVR_{distance,A}$ = 1.071, $RVR_{distance,V}$ = 1.318; $RVR_{size,A}$ = 1.019, $RVR_{size,V}$ = 1.236) allows the findings $\hat{D}_V > \hat{D}_A$ and $\hat{S}_V > \hat{S}_A$ to be transferred from virtual scenes to corresponding real scenes without the persistence of the aforesaid scene-specific exception.

4.2 Basic mode of perception

Regarding RQ 2, there is evidence that the basic mode of perception (processing of single- vs. multi-domain stimuli) as such alters perceptual estimates of geometric dimensions in virtual rooms. Mean estimates based on acoustic-visual stimuli did not equal the average of the mean estimates based on either only acoustic or only visual stimuli. Rather, mean estimates of source distance under the acoustic-visual condition (with acoustic-visually congruent stimuli) were located at 85% (music) of the range between the mean estimates of the levels **A** and **V**, mean estimates of room size at 92% (music) and 94% (speech), indicating that under the multi-domain condition visual information was weighted significantly higher than acoustic information. Though the distance estimation of the speech performance did not show a significant effect of perceptual mode, the mean estimates still accounted for 60% of the range between the mean estimates at levels **A** and **V**. When loading the mean estimates with the above-mentioned compensation factors, the percentages concerning music changed from 85% to 84% for source distance and from 92–86% for room size. Hence, the finding on RQ 2 may be transferred to reality in principle.

4.3 Properties of auralized and visualized rooms

Considering the multi-domain mode of perception and applying the conflicting stimulus paradigm, the distance and size estimates depended significantly on both the acoustic and the visual properties of the stimuli (RQs 3 and 4). Generally, about 89% of the explained variance arose from the entire visual and 10% from the entire acoustic information provided by the virtual environment. For both egocentric distance and room size perception, acoustic information showed a slightly greater proportion of explained variance under the speech than under the music condition.

In accordance with the MLE modeling of auditory-visual integration in principle, the acoustic and visual proportions of the explained variance appear to vary strongly according to the availability and, respectively, the richness of the cues in the particular domains: A preliminary experiment under substantially restricted visualization conditions (reduced field of view, reduced spatial resolution, still photographs instead of moving pictures, no maximal acoustic-visual congruency due to visible loudspeakers as sound sources) and non-restricted auralization conditions (identical auralization system) yielded a reversed order of proportions of the explained variance (cf. 2.7), which amounted to 33% for factor *Visualized room*, and 66% for factor *Auralized room* ([18], p. 392).

Against the background of the prevalent term *auditory-visual interaction* (or similar) it is remarkable, that at least no *statistical* interaction effect of the acoustic and the visual stimulus properties on egocentric source distance and room size perception was found that was significant (1.3, RQ5). Looking at perceived geometric dimensions as supramodal unified features specifying spatial notions, both acoustic and visual properties, and therefore both the auditory and the visual modalities, appear to contribute (regardless of variable weights) directly to the values of these features, and no interaction (non-additive) effects appear to complicate this straightforward principle. Hence, the modeling of auditory-visual integration of distance and room size perception will not have to include non-additive effects for the time being.

Since the involved modalities and the mode of perception were constant across all factor levels, it may be assumed that VR-induced biases apply likewise to all factor levels of the conflicting stimulus paradigm and their combinations. Hence, the findings on RQs 3 to 5, i.e., the inferential statistics and the η^2-based proportional accounts for the estimates, may be transferred from virtuality to reality in principle. At the descriptive level, the estimates might again be compensated for virtualization by loading them with $RVR_{distance,AV} = 1.284$ and $RVR_{size,AV} = 1.191$, respectively [38].

4.4 Complex independent variables and interfering factors

Within the test design, the presence and properties of the acoustic and visual domains were varied to experimentally dissociate the auditory and the visual modalities. Because this variation was categorical, i.e., comprising the entire *environmental* conditions of the scenes instead of either mere *distance* or mere *room size* cues, the results may be transferred to the perceptual modalities hearing and vision as such—at least for closed spaces, and within the boundaries of generalization given by the content types, rooms, and samples. Auditory-visual distance perception may in principle be influenced not only by physical distance, but by any structural (room size, room shape) and material properties that affect those acoustic cues (1.2) that are also affected by physical distance (cf. [124]). Since the domain proportions found in the present study cannot directly be compared to the weights determined in [79], which are based on mere distance-related cues, those interfering factors had to be experimentally dissociated and, where applicable, included in physical-perceptual models of auditory-visual distance perception.

4.5 Additional observations

There were some additional results on factors and measures which were not explicitly asked for by the RQs:

a. Both egocentric distance and room size mean estimates, regardless of whether based on acoustic, visual or acoustic-visual stimuli, were obviously lower for speech than for music (though this was not hypothesized or tested, see 2.7). Hence, there is a reason for hypothesizing an influence of content type. This might be due to differences between music and speech regarding, e.g., the bandwidth and energy distribution of the frequency spectra carrying spatial information, perceptual filtering and processing, receptiveness, and/or experiential geometric situations (non-mediatized speech is normally received from lower distances and within smaller rooms than non-mediatized music).

b. Both the non-significant interaction effect and the particular mean estimates in the experiment according to the conflicting stimulus paradigm indicated that acoustic-visual (mainly spatial) congruency of the stimulus properties did not lead to minimum, maximum or especially accurate mean estimates. This observation is not apt to constitute a general hypothesis, since congruency might play a greater role by contrast with a greater range of the incongruencies (e.g., further-away sound sources) or a greater number of incongruent properties (e.g., including incongruent content).

c. Egocentric distance mean estimates were most accurate under the acoustic-visual (music) and visual (speech) condition; the room size mean estimates, which were generally inaccurate, likely due to the lack of the visual rendering of the rooms' rear part, were most accurate under the acoustic condition. In contrast to previous studies [32, 36], regardless of general under- or overestimations of the geometric properties ($\hat{D}/D \neq 1$) under the acoustic-visual condition, neither an *increasing* underestimation nor an *increasing* overestimation was conspicuous, rather $\hat{D}/D \approx$ const.

d. Looking at the conflicting stimulus paradigm, the minimum and maximum mean estimates of both source distance and room size did not consistently correspond to the minimum and maximum physical distances and sizes. *Perceived source distance* and *perceived room size* were each influenced by the physical source distance, the physical room size and potentially other properties of the virtual scenes.

e. Because mean estimates based on purely acoustic stimuli were generally higher in low-absorbent than in high-absorbent rooms (cf. [18]), the range of mean estimates introduced by the factor *Domain* was also generally smaller in low-absorbent rooms. This caused the respective mean estimates under the acoustic condition to be more consistent with—and in the case of room size, even more accurate than—those under the visual and acoustic-visual conditions. Therefore, when visual information is unavailable, perception may exploit the greater amount of acoustic information provided by low-absorbent rooms to improve the accuracy of room size perception. Acoustic absorption may influence not only the values but also the availability and/or acuity of auditory cues (cf. 1.2).

Observations (d) and (e) and differences between the studies regarding domain proportions (4.3) give reason to hypothesize that structural and material properties of rooms influence distance perception. Thus, an additional experimental dissociation of the factors physical source distance, physical room size, and acoustic absorption (all else being equal) might be instructive. Furthermore, more detailed physical factors affecting both the acoustic and the visual domain might be disentangled (primary structures, secondary structures, materials). Because of the trade-off between the requirement of ecological stimulus validity and the costs of stimulus production, it might be worth investigating the moderating effects of certain aspects of virtualization (direct rendering, stereoscopy, visually moving persons). In the future, one major aim of research into the perception of geometric properties might be the connection of the modeling of internal mechanisms and the physical-perceptual modeling.

5. Conclusion

The influence of the presence as well as of the properties of acoustic and visual information on the perceived egocentric distance and room size was investigated applying both a co-presence and a conflicting stimulus paradigm. Constant music and speech renditions in six different rooms were presented using dynamic binaural synthesis and stereoscopic semi-panoramic video projection. Experimentation corroborated that perceptual mean estimates of geometric dimensions based on only visual information considerably exceeded those based on only acoustic information in general. However, the perceptual mode as such (single- vs. multi-domain stimuli) altered the perceptual estimates of geometric dimensions: Under the acoustic-visual condition with acoustic-visually congruent stimuli, the presence of visual geometric information was generally given more weight than the presence of acoustic information. While the egocentric distance estimation under the acoustic-visual condition did not tend to be compressed for music, it did for speech. When only acoustic stimuli were available, the greater amount of acoustic information provided by low-absorbent rooms appeared to be perceptually exploited to improve the accuracy of room size perception. Within the multi-domain mode of perception involving 30 acoustic-visually incongruent and 6 congruent stimuli, auditory-visual estimation of geometric dimensions in rooms relied about nine-tenths on the variation of visual, about one-tenth on the variation of acoustic properties, and negligibly on the interaction of the variation of the particular properties. Both the auditory and the visual sensory systems contribute to the perception of geometric dimensions in a straightforward manner. The observation of generally lower estimates for speech than for music needs to be corroborated and clarified. Further experimentation dissociating the factors source distance, room size, and acoustic absorption (all else being equal) is needed to clarify their particular influence on auditory-visual distance and room size perception.

Ethics statement

According to the funding institution (Deutsche Forschungsgemeinschaft) an ethical approval is not required, since the respective indications do not apply [125]. The study was conducted under the ethical principles of the appropriate national professional society (Deutsche Gesellschaft für Psychologie) [126].

Acknowledgements

This work was carried out as a part of the project "Audio-visual perception of acoustical environments", funded by the Deutsche Forschungsgemeinschaft (DFG MA 4343/1-1) within the framework of the research unit SEACEN, coordinated by Technische Universität Berlin and Rheinisch-Westfälische Technische Hochschule Aachen, Germany. We thank the staff of the performance rooms for their friendly cooperation, the Berlin Budapest Quartet (Dea Szücs, Éva Csermák, Itamar Ringel, Ditta Rohmann) and actress Ilka Teichmüller for their performances, Alexander Lindau, Fabian Brinkmann, and Vera Erbes for the in-situ acquisition of the rooms' acoustic and visual properties, Mina Fallahi for the geometric picture editing, Annika Natus, Alexander Haßkerl, and Shamir Ali-Khan for the 3D video shooting and post-production, and all test participants. Finally, we thank the two anonymous reviewers for critically reading the manuscript and suggesting substantial improvements.

Conflict of interest

The authors have no conflict of interest to declare.

A. Appendix

Measure	Content	A room	V room off	1	3	4	5	5	6
Mean	Music (n = 50)	off	—	10.87	6.85	10.25	10.52	11.80	15.47
		1	7.46	10.79	7.19	10.23	10.13	10.41	14.41
		2	7.18	10.51	7.32	10.28	11.07	10.60	13.85
		3	8.07	9.92	7.25	9.89	10.72	10.63	13.76
		4	6.74	10.63	7.44	9.84	10.17	10.01	13.12
		5	5.71	9.19	7.26	9.60	9.92	9.77	13.09
		6	12.91	11.16	8.43	11.24	11.22	11.61	15.23
	Speech (n = 38)	off	—	9.96	5.91	9.56	10.33	10.19	12.45
		1	7.10	8.88	6.11	8.82	8.87	9.37	11.52
		2	8.07	8.53	6.46	8.98	9.04	9.09	11.39
		3	6.67	8.62	6.00	8.69	8.72	8.67	11.26
		4	5.90	8.28	5.67	8.31	8.38	8.70	10.68
		5	5.30	7.67	6.03	8.12	8.30	8.03	10.62
		6	11.07	9.72	7.07	9.72	9.92	9.94	12.23

Measure	Content	A room	V room						
			off	1	3	4	5	5	6
STD	Music (n = 50)	off	—	4.36	3.04	3.53	2.59	3.54	3.61
		1	3.54	3.77	2.40	3.03	3.07	3.10	3.83
		2	4.03	3.78	2.74	3.11	3.01	2.94	3.63
		3	4.19	3.15	2.90	2.83	3.20	3.05	3.69
		4	3.09	4.24	2.69	2.77	3.03	2.88	4.42
		5	2.93	3.08	2.73	3.15	3.39	3.30	4.04
		6	4.83	4.12	3.90	3.57	3.58	4.17	3.72
	Speech (n = 38)	off	—	4.10	2.02	2.76	4.25	3.90	4.10
		1	3.52	3.15	2.28	2.74	2.67	2.95	4.44
		2	4.17	3.01	2.55	2.81	2.50	3.02	3.95
		3	3.04	3.46	2.12	2.73	2.36	2.44	4.48
		4	2.75	2.81	1.93	2.63	2.57	3.08	4.54
		5	3.34	3.21	2.55	2.38	2.91	2.67	4.83
		6	5.82	3.70	2.73	3.01	3.37	3.60	4.03

Table 11.
Descriptive statistics of perceived source distance ($\hat{D}$).

Measure	Content	A room	V room						
			off	1	3	4	5	5	6
Mean	Music (n = 50)	off	—	30.76	24.39	18.36	25.03	18.79	32.28
		1	16.89	29.38	22.66	19.48	22.94	19.32	30.06
		2	18.36	28.97	24.10	19.05	23.75	19.34	30.63
		3	14.99	29.95	22.94	18.96	22.22	18.97	29.94
		4	12.51	28.70	22.78	18.20	22.07	18.88	28.26
		5	10.56	27.02	22.66	17.99	22.45	18.11	28.13
		6	31.88	31.52	26.18	21.73	25.64	23.12	33.40
	Speech (n = 38)	off	—	29.29	21.56	17.18	21.41	17.43	32.04
		1	16.19	26.51	22.53	17.71	20.37	18.69	29.28
		2	20.75	27.78	23.43	19.07	20.69	19.34	30.03
		3	11.53	25.70	21.71	17.15	19.36	17.39	28.98
		4	9.85	25.07	20.89	16.29	19.54	17.00	26.69
		5	8.42	24.60	20.91	17.04	18.94	16.96	26.69
		6	31.93	29.47	26.04	20.85	23.25	21.30	33.11
STD	Music (n = 50)	off	—	7.01	7.09	6.67	7.45	6.79	8.76
		1	6.58	8.39	6.21	5.62	7.40	7.02	9.12
		2	7.05	8.89	7.96	6.50	7.10	6.66	8.73
		3	6.30	8.06	7.10	6.36	7.17	5.69	8.73
		4	6.36	7.88	7.02	5.49	6.74	6.57	9.00
		5	4.83	9.12	7.24	5.54	7.51	6.54	8.96
		6	8.67	7.91	7.91	8.58	8.21	8.68	9.20

Measure	Content	A room	V room						
			off	1	3	4	5	5	6
	Speech (n = 38)	off	—	9.68	7.31	5.97	9.02	7.43	9.43
		1	7.28	10.02	7.79	8.06	8.25	8.15	11.07
		2	7.56	9.79	8.60	7.53	8.09	8.02	11.42
		3	5.94	11.32	10.34	7.93	7.68	7.59	12.80
		4	3.48	11.35	9.37	6.92	8.40	7.91	12.47
		5	5.89	10.99	10.12	8.07	9.18	8.03	12.94
		6	8.39	9.36	8.82	10.39	9.82	9.66	10.71

Table 12.
Descriptive statistics of perceived room size $(\hat{S})$.

Author details

Hans-Joachim Maempel* and Michael Horn
Federal Institute for Music Research, Berlin, Germany

*Address all correspondence to: maempel@sim.spk-berlin.de

IntechOpen

References

[1] Cabrera D, Nguyen A, Choi YJ. Auditory versus visual spatial impression: A study of two auditoria. In: Barrass S, Vickers P, editors. Proc. of ICAD 04-Tenth Meeting of the Int. Conf. on Auditory Display. Sydney, Australia: International Community for Auditory Display (ICAD); 2004

[2] Kuusinen A, Lokki T. Auditory distance perception in concert halls and the origins of acoustic intimacy. Proceedings of the Institute of Acoustics. 2015;**37**(3):151-158

[3] Hyde JA. Discussion of the relation between initial time delay gap (ITDG) and acoustical intimacy: Leo Beranek's final thoughts on the subject, documented. Acoustics. 2019;**1**(3): 561-569. DOI: 10.3390/acoustics1030032

[4] Stevens JC, Marks LE. Cross-modality matching of brightness and loudness. Proceedings of the National Academy of Sciences of the USA. 1965;**2**: 407-411

[5] Thomas GJ. Experimental study of the influence of vision on sound localization. Journal of Experimental Psychology. 1941;**28**:163-177. DOI: 10.1037/h0055183

[6] Howard IP, Templeton WB. Human Spatial Orientation. London: Wiley; 1966

[7] Gardner MB. Proximity image effect in sound localization. The Journal of the Acoustical Society of America. 1968; **43**(1):163. DOI: 10.1121/1.1910747

[8] Mateeff S, Hohnsbein J, Noack T. Dynamic visual capture: Apparent auditory motion induced by a moving visual target. Perception. 1985;**14**: 721-727. DOI: 10.1068/p140721

[9] Kitajima N, Yamashita Y. Dynamic capture of sound motion in three-dimensional space. Perceptual and Motor Skills. 1999;**89**(3):1139-1158. DOI: 10.2466/pms.1999.89.3f.1139

[10] Kohlrausch A, van de Par S. Audio-visual interaction in the context of multimedia applications. In: Blauert J, editor. Communication Acoustics (Chapter 5). Berlin: Springer; 2005. pp. 109-138. DOI: 10.1007/3-540-27437-5_5

[11] Shams L, Kamitani Y, Shimojo S. Visual illusion induced by sound. Cognitive Brain Research. 2002;**14**: 147-152. DOI: 10.1016/S0926-6410(02) 00069-1

[12] Andersen TS, Tippana K, Sams M. Factors influencing audiovisual fission and fusion illusions. Cognitive Brain Research. 2004;**21**:301-308. DOI: 10.1016/j.cogbrainres.2004. 06.004

[13] Vatakis A, Spence C. Audiovisual synchrony perception for music, speech, and object actions. Brain Research. 2006;**1111**(1):134-142. DOI: 10.1016/j. brainres.2006.05.078

[14] MacDonald J, McGurk H. Visual influences on speech perception process. Perception & Psychophysics. 1978;**24**:253-257. DOI: 10.3758/ BF03206096

[15] Beerends JG, de Caluwe FE. The influence of video quality on perceived audio quality and vice versa. Journal of the Audio Engineering Society. 1999;**47**: 355-362

[16] Larsson P, Västfjäll D, Kleiner M. Auditory-visual interaction in real and virtual rooms. In: 3rd Convention of the EAA. Spain: Sevilla; 2002

[17] Larsson P, Väljamäe A. Auditory-visual perception of room size in virtual environments. In: Proc. of the 19th Int.

Congress on Acoustics. Madrid; 2007 PPA-03-001

[18] Maempel H-J, Jentsch M. Auditory and visual contribution to egocentric distance and room size perception. Building Acoustics. 2013;**20**(4):383-401. DOI: 10.1260/1351-010X.20.4.383

[19] Treisman A. The binding problem. Current Opinion in Neurobiology. 1996; **6**(2):171-178. DOI: 10.1016/S0959-4388 (96)80070-5

[20] Bishop ID, Rohrmann B. Subjective responses to simulated and real environments: A comparison. Landscape and Urban Planning. 2003; **65**(4):261-277. DOI: 10.1016/S0169-2046(03)00070-7

[21] de Kort YAW, IJsselsteijn WA, Kooijman J, Schuurmans Y. Virtual laboratories: Comparability of real and virtual environments for environmental psychology. Presence—Teleoperators and Virtual Environments. 2003;**12**(4): 360-373. DOI: 10.1162/ 105474603322391604

[22] Billger M, Heldal I, Stahre B, Renstrom K. Perception of color and space in virtual reality: a comparison between a real room and virtual reality models. In: Rogowitz BE, Pappas TN, editors. Proc. of SPIE, Human Vision and Electronic Imaging IX, San Jose, California, USA. Vol. 5292. Bellingham, WA: Society of Photographic Instrumentation Engineers (SPIE); 2004. pp. 90-98. DOI: 10.1117/12.526986

[23] Kuliga SF, Thrash T, Dalton RC, Hölscher C. Virtual reality as an empirical research tool – Exploring user experience in a real building and a corresponding virtual model. Computers, Environment and Urban Systems. 2015;**54**:363-375. DOI: 10.1016/ j.compenvurbsys.2015.09.006

[24] Nielsen SH. Auditory distance perception in different rooms. Journal of the Audio Engineering Society. 1993;**41**: 755-770

[25] Bronkhorst AW, Houtgast T. Auditory distance perception in rooms. Nature. 1999;**397**:517-520. DOI: 10.1038/ 17374

[26] Bronkhorst AW, Zahorik P. The direct-to-reverberant ratio as cue for distance perception in rooms. The Journal of the Acoustical Society of America. 2002;**111**(5):2440-2441. DOI: 10.1121/1.4809156

[27] Loomis JM, Klatzky RL, Golledge RG. Auditory distance perception in real, virtual, and mixed environments. In: Ohta Y, Tamura H, editors. Mixed Reality: Merging Real and Virtual Worlds. Ohmsha: Tokyo; 1999. pp. 201-214

[28] Zahorik P, Brungart DS, Bronkhorst AW. Auditory distance perception in humans: A summary of past and present research. Acta Acustica united with Acustica. 2005;**91**(3): 409-420

[29] Kolarik AJ, Cirstea S, Pardhan S. Discrimination of virtual auditory distance using level and direct-to-reverberant ratio cues. The Journal of the Acoustical Society of America. 2013; **134**(5):3395. DOI: 10.1121/1.4824395

[30] Kolarik AJ, Moore BCJ, Zahorik P, Cirstea S, Pardhan S. Auditory distance perception in humans: A review of cues, development, neuronal bases, and effects of sensory loss. Attention, Perception, & Psychophysics. 2016;**78**: 373-395. DOI: 10.3758/s13414-015-1015-1

[31] Moulin S, Nicol R, Gros L. Auditory distance perception in real and virtual environments. In: SAP '13, Proc. of the ACM Symposium on Applied Perception. Dublin: ACM; 2013. p. 117. DOI: 10.1145/2492494.2501876

[32] Kearney G, Gorzel M, Boland F, Rice H. Depth perception in interactive virtual acoustic environments using higher order ambisonic sound fields. In: 2nd Int. Symposium on Ambisonics and Spherical Acoustics. Berlin: Univ.-Verl. TU; 2010

[33] Calcagno ER, Abregu EL, Eguia MC, Vergara R. The role of vision in auditory distance perception. Perception. 2012; **41**(2):175-192. DOI: 10.1068/p7153

[34] Zahorik P. Assessing auditory distance perception using virtual acoustics. The Journal of the Acoustical Society of America. 2002;**111**(4): 1832-1846. DOI: 10.1121/1.1458027

[35] Chan JS, Lisiecka D, Ennis C, O'Sullivan C, Newell FN. Comparing audiovisual distance perception in various 'real' and 'virtual' environments. Perception ECVP Abstract. 2009;**38**:30

[36] Rébillat M, Boutillon X, Corteel È, Katz BFG. Audio, visual, and audio-visual egocentric distance perception in virtual environments. In: EAA Forum Acusticum 2011. Denmark: Aalborg; 2011. pp. 482-487

[37] Rébillat M, Boutillon X, Corteel È, Katz BFG. Audio, visual, and audio-visual egocentric distance perception by moving subjects in virtual environments. ACM Transactions on Applied Perception. 2012;**9**:19:1-19:17. DOI: 10.1145/2355598.2355602

[38] Maempel H-J, Horn M. Audiovisual perception of real and virtual rooms. Journal of Virtual Reality and Broadcasting. 2017;**14**(5):1-15. DOI: 10.20385/1860-2037/14.2017.5

[39] Cabrera D, Jeong D, Kwak HJ, Kim J-Y. Auditory room size perception for modelled and measured rooms. In: Internoise, the 2005 Congress and Exposition on Noise Control Engineering. Rio de Janeiro, Brazil: Institute of Noise Control Engineering - USA (INCE-USA); 2005

[40] Cabrera D, Pop C, Jeong D. Auditory room size perception: a comparison of real versus binaural sound-fields. In: Acoustics. Christchurch, New Zealand; 2006. pp. 417-422

[41] Cabrera D. Acoustic clarity and auditory room size perception. In: 14th Int. Congress on Sound & Vibration. Cairns, Australia; 2007

[42] Hameed S, Pakarinen J, Valde K, Pulkki V. Psychoacoustic cues in room size perception. In: AES 116th Convention. Berlin, Germany: Audio Engineering Society; 2004. Convention Paper 6084

[43] Yadav M, Cabrera D, Martens WL. Auditory room size perceived from a room acoustic simulation with autophonic stimuli. Acoustics Australia. 2011;**39**(3):101-105

[44] Cutting JE, Vishton PM. Perceiving layout and knowing distances: the integration, relative potency, and contextual use of different information about depth. In: Epstein W, Rogers S, editors. Perception of Space and Motion (Chapter 3). San Diego et al: Academic Press; 1995. pp. 69-117. DOI: 10.1016/B978-012240530-3/50005-5

[45] Mehrabi M, Peek EM, Wuensche BC, Lutteroth C. Making 3D work: A classification of visual depth cues, 3D display technologies and their applications. Proc. of the 14th Australasian User Interface conference (AUIC 2013), Adelaide, Australia. Conferences in Research and Practice in Information Technology (CRPIT). 2013; **139**:91-100. DOI: 10.5555/2525493.2525503

[46] Renner RS, Velichkovsky BM, Helmert JR. The perception of egocentric distances in virtual environments—A review. ACM

Computing Surveys. 2013;**46**(2):1-40. DOI: 10.1145/2543581.2543590

[47] Zahorik P. Audio/visual interaction in the perception of sound source distance. In: Ochmann M, Vorländer M, Fels J, editors. ICA 2019 Aachen. Proc. of the 23rd Int. Congress on Acoustics, Aachen, Germany. Berlin: Deutsche Gesellschaft für Akustik; 2019. pp. 7927-7931

[48] Loomis JM, Da Silva JA, Philbeck JW, Fukusima SS. Visual perception of location and distance. Current Directions in Psychological Science. 1996;**5**(3):72-77. DOI: 10.1111/1467-8721.ep10772783

[49] Loomis JM, Knapp JM. Visual perception of egocentric distance in real and virtual environments. In: Hettinger LJ, Haas MW, editors. Virtual and Adaptive Environments. Applications, Implications, and Human Performance Issues. Mahwah, NJ: Erlbaum; 2003. pp. 21-46

[50] Plumert JM, Kearney JK, Cremer JF, Recker K. Distance perception in real and virtual environments. ACM Transactions on Applied Perception. 2005;**2**(3):216-233. DOI: 10.1145/1077399.1077402

[51] Armbrüster C, Wolter M, Kuhlen T, Spijkers W, Fimm B. Depth perception in virtual reality: Distance estimations in peri- and extrapersonal space. CyberPsychology & Behavior. 2008; **11**(1):9-15. DOI: 10.1089/cpb.2007.9935

[52] Klein E, Swan JE, Schmidt GS, Livingston MA, Staadt OG. Measurement protocols for medium-field distance perception in large-screen immersive displays. In: IEEE Virtual Reality. Vol. 2009. Lafayette, Lousiana, USA; 2009. pp. 107-113. DOI: 10.1109/VR.2009.4811007

[53] Naceri A, Chellali R, Dionnet F, Toma S. Depth perception within virtual environments: A comparative study between wide screen stereoscopic displays and head mounted devices. In: 2009 Computation World: Future Computing, Service Computation, Cognitive, Adaptive, Content, Patterns. 2009. pp. 460-466. DOI: 10.1109/ComputationWorld.2009.91

[54] Ziemer CJ, Plumert JM, Cremer JF, Kearney JK. Estimating distance in real and virtual environments: Does order make a difference? Attention, Perception, & Psychophysics. 2009;**71**(5):1095-1106. DOI: 10.3758/APP.71.5.1096

[55] Alexandrova IV, Teneva PT, de la Rosa S, Kloos U, Bülthoff HH, Mohler BJ. Egocentric distance judgments in a large screen display immersive virtual environment. In: 7th Symposium on Applied Perception in Graphics and Visualization. Los Angeles, CA, USA; 2010. pp. 57-60. DOI: 10.1145/1836248.1836258

[56] Interrante V, Anderson L, Ries B. Distance perception in immersive virtual environments, revisited. In: Proc. of the IEEE Virtual Reality Conf. (VR'06). New York: IEEE; 2006. pp. 3-10. DOI: 10.1109/VR.2006.52

[57] Bruder G, Argelaguet F, Olivier AH, Lécuyer A. CAVE size matters: Effects of screen distance and parallax on distance estimation in large immersive display setups. Presence—Teleoperators and Virtual Environments. 2016;**25**(1): 1-16. DOI: 10.1162/PRES_a_00241

[58] Gadia D, Galmonte A, Agostini T, Viale A, Marini D. Depth and distance perception in a curved, large screen virtual reality installation. In: Woods AJ, Holliman NS, Merritt JO, editors. Stereoscopic Displays and Applications XX. Proc. of SPIE. 2009;**7237**:723718. DOI: 10.1117/12.805809

[59] Creem-Regehr SH, Willemsen P, Gooch AA, Thompson WB. The

influence of restricted viewing conditions on egocentric distance perception: Implications for real and virtual indoor environments. Perception. 2005;**34**(2):191-204. DOI: 10.1068/p5144

[60] Kruszielski LF, Kamekawa T, Marui A. The influence of camera focal length in the direct-to-reverb ratio suitability and its effect in the perception of distance for a motion picture. In: AES 131st Convention. New York; 2011. Convention paper 8580.

[61] Anderson PW, Zahorik P. Auditory/visual distance estimation. Accuracy and variability. Front Psychol. 2014;**5**: 1097. DOI: 10.3389/fpsyg.2014.01097

[62] Larsson P, Västfjäll D, Kleiner M. Ecological acoustics the multimodal perception of rooms – real and unreal experiences of auditory-visual virtual environments. In: Hiipakka J, Zacharov N, Takala T, editors. 2001 Int. Conf. on Auditory Display. Espoo, Finland: Helsinki University of Technology; 2001

[63] Thurlow WR, Jack CE. Certain determinants of the "ventriloquism effect". Perceptual and Motor Skills. 1973;**36**(suppl. 3):1171-1184. DOI: 10.2466/pms.1973.36.3c.1171

[64] Jack CE, Thurlow WR. Effects of degree of visual association and angle of displacement on the "ventriloquism" effect. Perceptual and Motor Skills. 1973;**37**(3):967-979. DOI: 10.1177/003151257303700360

[65] Chen L, Vroomen J. Intersensory binding across space and time. A tutorial review. Attention, Perception, & Psychophysics. 2013;**75**(5):790-811. DOI: 10.3758/s13414-013-0475-4

[66] Mershon DH, Desaulniers DH, Amerson TLJ, Kiefer SA. Visual capture in auditory distance perception. Proximity image effect reconsidered.

The Journal of Auditory Research. 1980; **20**:129-136

[67] Zahorik P. Estimating sound source distance with and without vision. Optometry and Vision Science. 2001; **78**(5):270-275

[68] Côté N, Koehl V, Paquier M. Ventriloquism effect on distance auditory cues. In: Acoustics 2012 Joint Congress (11ème Congrès Français d'Acoustique—2012 Annual IOA Meeting), Apr 2012. France: Nantes; 2012. pp. 1063-1067

[69] Postma BNJ, Katz BFG. The influence of visual distance on the room-acoustic experience of auralizations. The Journal of the Acoustical Society of America. 2017; **142**(5):3035-3046. DOI: 10.1121/1.5009554

[70] Ernst MO, Banks MS. Humans integrate visual and haptic information in a statistically optimal fashion. Nature. 2002;**415**:429-433. DOI: 10.1038/415429a

[71] Battaglia PW, Jacobs RA, Aslin RN. Bayesian integration of visual and auditory signals for spatial localization. Journal of the Optical Society of America. A. 2003;**20**(7):1391-1397. DOI: 10.1364/josaa.20.001391

[72] Alais D, Burr D. The ventriloquist effect results from near-optimal bimodal integration. Current Biology. 2004;**14**(3):257-262. DOI: 10.1016/j.cub.2004.01.029

[73] Finnegan DJ, Proulx MJ, O'Neill E. Compensating for distance compression in audiovisual virtual environments using incongruence. In: Proc. of the 2016 CHI Conf. on Human Factors in Computing Systems (CHI'16). New York: ACM; 2016. pp. 200-212. DOI: 10.1145/2858036.2858065

[74] Agganis BT, Muday JA, Schirillo JA. Visual biasing of auditory localization in azimuth and depth. Perceptual and Motor

Skills. 2010;**111**(3):872-892. DOI: 10.2466/22.24.27.PMS.111.6.872-892

[75] Hládek L, Le Dantec CC, Kopčo N, Seitz A. Ventriloquism effect and aftereffect in the distance dimension. Proceedings of Meetings on Acoustics. 2013;**19**:050042. DOI: 10.1121/1.4799881

[76] Roach NW, Heron J, McGraw PV. Resolving multisensory conflict: A strategy for balancing the costs and benefits of audio-visual integration. Proceedings of the Royal Society of London. Series B. 2006;**273**(1598): 2159-2168. DOI: 10.1098/rspb.2006.3578

[77] Meijer D, Veselič S, Calafiore C, Noppeney U. Integration of audiovisual spatial signals is not consistent with maximum likelihood estimation. Cortex. 2019;**119**:74-88. DOI: 10.1016/j.cortex.2019.03.026

[78] André CR, Corteel E, Embrechts JJ, Verly JG, Katz BFG. Subjective evaluation of the audiovisual spatial congruence in the case of stereoscopic-3D video and wave field synthesis. International Journal of Human-Computer Studies. 2014;**72**(1):23-32. DOI: 10.1016/j.ijhcs.2013.09.004

[79] Mendonça C, Mandelli P, Pulkki V. Modelling the perception of audiovisual distance. Bayesian causal inference and other models. PLoS One. 2016;**11**(12): e0165391. DOI: 10.1371/journal.pone.0165391

[80] Spence C. Crossmodal correspondences: A tutorial review. Attention, Perception, & Psychophysics. 2011;**73**:971-995. DOI: 10.3758/s13414-010-0073-7

[81] Maempel H-J. Apples and oranges: A methodological framework for basic research into audiovisual perception. In: Hohmaier S, editor. Jahrbuch des Staatl. Inst. für Musikforschung 2016. Mainz et al: Schott; 2019. pp. 361–377. DOI: 10.14279/depositonce-6424.2

[82] Berg J, Rumsey F. Verification and correlation of attributes used for describing the spatial quality of reproduced sound. In: AES 19th Int. Conf., Schloss Elmau, Germany. Article No. 1932. New York City, New York, USA: Audio Engineering Society; 2001

[83] Rumsey F. Spatial quality evaluation for reproduced sound: Terminology, meaning, and a scene-based paradigm. Journal of the Audio Engineering Society. 2002;**50**(9): 651-666

[84] Bizley JK, Maddox RK, Lee AKC. Defining auditory-visual objects: Behavioral tests and physiological mechanisms. Trends in Neurosciences. 2016;**39**(2):74-85. DOI: 10.1016/j.tins.2015.12.007

[85] Faul F, Erdfelder E, Lang A-G, Buchner A. G*power 3: A flexible statistical power analysis program for the social, behavioral, and biomedical sciences. Behavior Research Methods. 2007;**39**:175-191. DOI: 10.3758/BF03193146

[86] Rasch B, Friese M, Hofmann W, Naumann E. Quantitative Methoden. 3rd ed. Vol. 2. Heidelberg: Springer; 2010

[87] Deutsches Institut für Normung e.V. DIN EN ISO 3382-1 Akustik – Messung von Parametern der Raumakustik – Teil 1: Aufführungsräume. Berlin: Beuth; 2009

[88] Beranek L. Concert hall acoustics. Journal of the Audio Engineering Society. 2008;**56**(7/8):532-544

[89] Harris CM. Absorption of sound in air versus humidity and temperature. The Journal of the Acoustical Society of America. 1966;**40**(1):148-159. DOI: 10.1121/1.1910031

[90] Lindau A, Schultz F, Horn M, Brinkmann F, Erbes V, Fuß A, Maempel H-J, Weinzierl S. Raumakustische Messungen in sechs Aufführungsräumen: Konzerthaus/ Kleiner Saal (Berlin), Jesus-Christus-Kirche (Berlin), Kloster Eberbach/ Basilika (Eltville am Rhein), Renaissance-Theater (Berlin), Komische Oper (Berlin), Gewandhaus/Großer Saal (Leipzig), measurement reports (research data). 2021:i-VI/16. DOI: 10.14279/depositonce-11947

[91] Lindau A, Weinzierl S. FABIAN – An instrument for software-based measurement of binaural room impulse responses in multiple degrees of freedom. In: 24. Leipzig: Tonmeistertagung; 2006

[92] Lindau A, Hohn T, Weinzierl S. Binaural resynthesis for comparative studies of acoustical environments. In: AES 122nd Convention. Vienna; 2007. Preprint 7032

[93] Hendrickx E, Stitt P, Messonier J-C, Lyzwa J-M, Katz BFG, de Boishéraud C. Influence of head tracking on the externalization of speech stimuli for non-individualized binaural synthesis. The Journal of the Acoustical Society of America. 2017;**141**(3):2011-2023. DOI: 10.1121/1.4978612

[94] McAnally KI, Martin RL. Sound localization with head movement: Implications for 3-d audio displays. Frontiers in Neuroscience. 2014;**8**(210): 1-6. DOI: 10.3389/fnins.2014.00210

[95] Lindau A, Maempel H-J, Weinzierl S. Minimum BRIR grid resolution for dynamic binaural synthesis. In: Forum Acusticum, European Acoustics Association. Proc. of Acoustics'08, Conference: Paris. Stuttgart: Hirzel; 2008. pp. 3851-3856

[96] Schultz F, Lindau A, Weinzierl S. Just noticeable BRIR grid resolution for

lateral head movements. In: DAGA 2009. Rotterdam; 2009. pp. 200-201

[97] Erbes V, Schultz F, Lindau A, Weinzierl S. An extraaural headphone system for optimized binaural reproduction. In: DAGA 2012, Darmstadt. Berlin: Deutsche Gesellschaft für Akustik; 2012. pp. 313-314

[98] Lindau A, Brinkmann F. Perceptual evaluation of headphone compensation in binaural synthesis based on non-individual recordings. Journal of the Audio Engineering Society. 2012;**60**(1/2):54-62

[99] Lindau A. The perception of system latency in dynamic binaural synthesis. In: DAGA 2009, Rotterdam. Berlin: Deutsche Gesellschaft für Akustik; 2009. pp. 1063-1066

[100] Lindau A, Estrella J, Weinzierl S. Individualization of dynamic binaural synthesis by real time manipulation of the ITD. In: AES 128th Convention. London: Audio Engineering Society; 2010. Preprint 8088

[101] Neidhardt A, Reif B. Minimum BRIR grid resolution for interactive position changes in dynamic binaural synthesis. In: Proc. of the 148th Int. AES Convention, Vienna, Austria. New York City, New York, USA: Audio Engineering Society; 2020. Available from: https://www.aes.org/e-lib/browse.cfm?elib=20788

[102] Werner S, Klein F, Neidhardt A, Sloma U, Schneiderwind C, Brandenburg K. Creation of auditory augmented reality using a position-dynamic binaural synthesis system— Technical components, psychoacoustic needs, and perceptual evaluation. Applied Sciences. 2021;**11**(3):1150. DOI: 10.3390/app11031150

[103] Speigle JM, Loomis JM. Auditory distance perception by translating

observers. In: Proc. of 1993 IEEE Research Properties in Virtual Reality Symposium. New York: IEEE; 1993. pp. 92-99. DOI: 10.1109/VRAIS.1993. 378257

[104] Genzel D, Schutte M, Brimijoin WO, MacNeilage PR, Wiegrebe L. Psychophysical evidence for auditory motion parallax. PNAS. 2018;**115**(16):4264-4269. DOI: 10.1073/ pnas.1712058115

[105] Carlile S, Leung J. The perception of auditory motion. Trends in Hearing. 2016;**20**:1-19. DOI: 10.1177/ 2331216516644254

[106] Rummukainen OS, Schlecht SJ, Habets EAP. Self-translation induced minimum audible angle. The Journal of the Acoustical Society of America. 2018; **144**(4):EL340–EL345. DOI: 10.1121/ 1.5064957

[107] Rosenberg LB. The effect of interocular distance upon operator performance using stereoscopic displays to perform virtual depth tasks. In: Proceedings of IEEE Virtual Reality Annual International Symposium. New York: IEEE; 1993. pp. 27-32. DOI: 10.1109/VRAIS.1993.380802

[108] Utsumi A, Milgram P, Takemura H, Kishino F. Investigation of errors in perception of stereoscopically presented virtual object locations in real display space. Proceedings of the Human Factors and Ergonomics Society Annual Meeting. 1994;**38**(4):250-254. DOI: 10.1177/154193129403800413

[109] Best S. Perceptual and oculomotor implications of interpupillary distance settings on a head-mounted virtual display. Proceedings of Naecon IEEE Nat. 1996;**1**:429-434. DOI: 10.1109/ NAECON.1996.517685

[110] Drascic D, Milgram P. Perceptual issues in augmented reality. In: Proc. SPIE 2653, Stereoscopic Displays and

Virtual Reality Systems III. 1996. pp. 123-134. DOI: 10.1117/12.237425

[111] Wartell Z, Hodges LF, Ribarsky W. Balancing fusion, image depth and distortion in stereoscopic head-tracked displays. In: SIGGRAPH '99: Proc. of the 26th annual conference on Computer graphics and interactive techniques. 1999. pp. 351-358. DOI: 10.1145/ 311535.311587

[112] Renner RS, Steindecker E, Müller M, Velichkovsky BM, Stelzer R, Pannasch S, et al. The influence of the stereo base on blind and sighted reaches in a virtual environment. ACM Transactions on Applied Perception. 2015;**12**(2):1-18. DOI: 10.1145/2724716

[113] Kim N-G. Independence of size and distance in binocular vision. Frontiers in Psychology. 2018;**9**:1-18. DOI: 10.3389/ fpsyg.2018.00988

[114] Willemsen P, Gooch AA, Thompson WB, Creem-Regehr SH. Effects of stereo viewing conditions on distance perception in virtual environments. Presence—Teleoperators and Virtual Environments. 2008;**17**(1): 91-101. DOI: 10.1162/pres.17.1.91

[115] Bruder G, Pusch A, Steinicke F. Analyzing effects of geometric rendering parameters on size and distance estimation in on-axis stereographics. In: SAP '12: Proc. of the ACM Symposium on Applied Perception. 2012. pp. 111-118. DOI: 10.1145/2338676.2338699

[116] Deutsches Institut für Normung e. V. DIN 33402–2 Ergonomie – Körpermaße Des Menschen – Teil 2: Werte. Berlin: Beuth; 2020

[117] Knapp JM, Loomis JM. Limited field of view of head-mounted displays is not the cause of distance underestimation in virtual environments. Presence—Teleoperators and Virtual Environments. 2004;**13**(5):

572-577. DOI: 10.1162/105474604254 5238

[118] Maempel HJ, Horn M. The virtual concert hall – A research tool for the experimental investigation of audiovisual room perception. International Journal on Stereoscopic and Immersive Media. 2017;**1**(1):78-98

[119] Lindau A, Weinzierl S. Assessing the plausibility of virtual acoustic environments. Acta Acustica United with Acustica. 2012;**98**(5):804-810. DOI: 10.3813/AAA.918562

[120] Olejnik S, Algina J. Generalized eta and omega squared statistics: Measures of effect size for some common research designs. Psychological Methods. 2003; **8**(4):434-447. DOI: 10.1037/1082-989X. 8.4.434

[121] Bakeman R. Recommended effect size statistics for repeated measures designs. Behavior Research Methods. 2005;**37**(3):379-384. DOI: 10.3758/ BF03192707

[122] Cohen J. Statistical Power Analysis for the Behavioral Sciences. 2nd ed. New Jersey: Erlbaum; 1988

[123] Lakens D. Calculating and reporting effect sizes to facilitate cumulative science: A practical primer for *t*-tests and ANOVAs. Frontiers in Psychology. 2013;**4**(863):1-12. DOI: 10.3389/fpsyg.2013.00863

[124] Cabrera D. Control of perceived room size using simple binaural technology. In: Martens WL, editor. Proc. of the 13th Int. Conf. on Auditory Display. Montréal, Canada: International Community for Auditory Display; 2007

[125] FAQ: Humanities and Social Sciences. Statement by an Ethics Committee [Internet]. Available from: https://www.dfg.de/en/research_fund ing/faq/faq_humanities_social_science/ index.html [Accessed: January 10, 2022]

[126] Berufsethische Richtlinien [Internet]. In: Berufsverband Deutscher Psychologinnen und Psychologen e.V. Deutsche Gesellschaft für Psychologie e.V. 2016. Available from: https://www. dgps.de/fileadmin/user_upload/PDF/ berufsethik-foederation-2016.pdf [Accessed: January 10, 2022]

Chapter 5

Reverberation and its Binaural Reproduction: The Trade-off between Computational Efficiency and Perceived Quality

Isaac Engel and Lorenzo Picinali

Abstract

Accurately rendering reverberation is critical to produce realistic binaural audio, particularly in augmented reality applications where virtual objects must blend in seamlessly with real ones. However, rigorously simulating sound waves interacting with the auralised space can be computationally costly, sometimes to the point of being unfeasible in real time applications on resource-limited mobile platforms. Luckily, knowledge of auditory perception can be leveraged to make computational savings without compromising quality. This chapter reviews different approaches and methods for rendering binaural reverberation efficiently, focusing specifically on Ambisonics-based techniques aimed at reducing the spatial resolution of late reverberation components. Potential future research directions in this area are also discussed.

Keywords: binaural audio, reverberation, Auralisation, Ambisonics, perceptual evaluation

1. Introduction

Reverberation results from pairing a sound source with an acoustic space. After emanating from the source, a sound wave will interact with its environment, undergoing reflection, diffraction and absorption. Thus, a listener will receive filtered replicas of the original wavefront (echoes) arriving from various directions at different times, causing the impression that the original sound persists in time. According to the so-called precedence effect, the direct sound allows a listener to determine the position of the sound source, while early reflections are generally not perceived as distinct auditory events [1–3]. As stated by Wallach et al. [1], the maximum delay after which a reflection is no longer 'fused' with the direct sound depends on the signal, being around 5 ms for single clicks and as long as 40 ms for complex signals such as speech or music [4]. Nevertheless, early reflections can broaden the perceived width of the source and shift its apparent position, as shown experimentally by Olive and Toole [5]. Furthermore, they can modify the signal's spectrum due to phase cancellation and subsequent comb filtering, as shown by Bech in his study on small-room acoustics [6]. Such phenomena can alter the perception of the room on a higher level. For example, Barron and Marshall [7] argued that the timing, direction of arrival, and spectra of early lateral reflections contribute to the sense of 'envelopment'—defined as the 'subjective impression of being surrounded by the sound'. The time delay between the direct sound and the first distinct echo has also been shown to be a relevant feature: in the case of small rooms, Kaplanis et al. [8] found that it was correlated with the perception of environment dimensions and 'presence'—or 'sense of being inside an enclosed space and feeling its boundaries'—while in the case of concert halls, Beranek [9] linked it to a sense of 'intimacy'.

As time passes and the sound waves that emanated from the source continue interacting with the environment, the temporal density of echoes increases, and the resulting sound field becomes more diffuse. At this point, temporal and spatial features of individual echoes become less relevant, and late reverberation can be characterised as a stochastic process. An important parameter used to define such process is the reverberation time (RT), or the 'duration required for the space-averaged sound energy density in an enclosure to decrease by 60 dB after the source emission has stopped' [10], which is generally proportional to the volume of the room. Yadav et al. [11] suggested that RT contributes to the perception of environment dimensions most significantly in large spaces, whereas early reflections have greater importance in small rooms. Although late reverberation is often modelled as diffuse and isotropic (i.e., with an even distribution of energy across directions from the listeners' point of view). Alary et al. [12] showed that this assumption may not always hold and directionality should be taken into account, especially for asymmetrical spaces, such as a corridor.

When reproduced binaurally (e.g., through headphones), it has been shown that reverberation increases the sense of externalisation, i.e., the illusion of virtual sound sources being outside the head, when compared to anechoic sounds [13, 14]. It has been suggested that this effect can be achieved even by just adding the early reflections [13], while the contribution of late reverberation (>80 ms) is smaller in comparison [15]. Previous studies have looked into the contribution of both monaural and binaural cues to the externalisation of reverberant binaural signals. Monaural cues have been shown to have limited importance by Hassager et al. [16] and Jiang et al. [17], who argued that spectral detail is not as critical in the reverberant sound as it is in the direct sound. Regardless, it has been reported that applying spectral correction (headphone equalisation) to binaural signals could increase externalisation and other subjective attributes when employing headphones with limited reproduction bandwidth [18, 19]. Binaural cues, on the other hand, have been shown to be critical: Leclere et al. [14] suggested that reverberation increases externalisation of a binaural signal as long as interaural differences are introduced. This is supported by Catic et al. [15], who reported a considerable decrease in externalisation when the reverberant part of auralised speech was presented diotically. Such effects have been linked to specific binaural cues, such as interaural level differences (ILDs) and interaural coherence (IC). Recent studies have reported correlations between the level of externalisation and the amount of temporal fluctuations of ILDs and IC in the binaural signals [14, 15, 20]. Moreover, Li et al. [21, 22] highlighted the importance of reverberation specifically in the contralateral ear signal, showing a stronger contribution to externalisation than its ipsilateral counterpart, which is explained by the fact that reverberation is proportionally louder on the contralateral side due to the head shadow effect. Finally, according to the 'room divergence effect', externalisation of simulated binaural signals increases when the rendered reverberation matches the listener's expectations given their prior knowledge of the room [23–25]. Head movements and vision also play an important role in spatial audio perception [26], but they are not covered here—for a thorough review on sound externalisation, the reader is referred to Best et al. [27].

In summary, reverberation greatly influences how a listener perceives an auditory scene by providing information on the room characteristics, the size and location of the sound sources and, in the case of binaural simulations, affecting the level of externalisation. Consequently, it should be modelled carefully when producing realistic acoustic simulations, although this can prove to be a challenging task in real-time systems with limited resources. The next sections of this chapter will address the issue of balancing computational efficiency and perceptual quality when simulating reverberation.

2. Simulating reverberation efficiently

Simulating reverberation can be useful in various applications. In some cases, such as music production, it has mainly an aesthetic value and may not require highly realistic simulations. In other cases, such as architectural acoustics, augmented reality (AR) and, to a lesser extent, virtual reality (VR), the goal is to recreate a real acoustic space, so reverberation needs to be modelled with sufficient accuracy. For instance, an AR system allows the users to perceive the real world integrated with a virtual layer, e.g., a videoconferencing application in which users, wearing a pair of AR glasses, see holograms of their interlocutors which look and sound as if they were in the same room. From the acoustic point of view, this is particularly challenging to implement because the listener is exposed to real sound sources as well as virtual ones, so the simulated acoustics should be realistic enough for the virtual and real sources to be appropriately blended. Even though highly realistic reverberation is often desired, it can easily become too expensive to simulate in real time for interactive applications, where the auditory scene is expected to vary over time—even more so if many virtual sources are simulated [28]. Therefore, it is relevant to explore simplified reverberation models that reduce computational costs without compromising quality.

In the most general case, reverberation is rendered by convolving a dry audio signal with a room impulse response (RIR), which is the time-domain acoustic transfer function between a sound source and a receiver in a given acoustic space (room), assuming that the system formed by these is linear and time-invariant. The RIR can be either measured acoustically [29] or in a simulated environment. Several simulation techniques have been proposed, which range from rigorous but computationally expensive physical models, such as the finite-difference time-domain method [30], to simpler but less accurate geometrical models, such as the image source method [31] or scattering delay networks [32]. Ray-tracing and cone-tracing are also popular techniques that allow for a variable degree of accuracy [28, 33–35], albeit the computational requirements can become rather intensive when sound sources move in space, and real-time implementations are often limited to very simplified models and/or renderings.

Reverberation may also be generated through computationally lighter 'convolution-less' methods, such as Schroeder reverberators [36] or feedback delay networks (FDN) [37–39]. Such techniques are generally less accurate than convolution-based methods but can be useful to efficiently model the less critical parts of the RIR such as the late-reverberation tail [40].

With the goal of finding a balance between computational cost and perceived quality, several parametric reverberation models have been proposed [40–47]. Most of them aim to alleviate computational costs by rendering early reflections with a higher temporal and spatial accuracy than late reverberation, based on the concept of mixing time, i.e., the instant after which the RIR does not perceivably change across different listeners' positions or orientations within the room (see **Figure 1**) [48].

An early example of this approach, known as 'hybrid' reverberation, was presented by Murphy and Stewart [40], who proposed to employ convolution-based rendering for early reflections and simpler methods (e.g., FDN) to produce late reverberation. A key aspect of the hybrid model is correctly establishing the mixing time, which depends on the room volume, being higher for larger rooms [48].

In spatial audio applications, it is important to accurately simulate the direction of arrival of early reflections (and of late reverberation, to a lesser extent) which adds yet another layer of difficulty to the process. This also means that the reproduction method should be able to replicate such spatial cues. An example of a playback system would be a loudspeaker array surrounding the listener that can simulate virtual sources and reflections through amplitude panning [49] or Ambisonics [50]. In the case of binaural audio, such systems may be mimicked through virtual loudspeakers, but other methods also exist, as discussed in Section 2.1.

Note that the scope of this chapter covers reverberation's spatial features from the listener's point of view, but not from the source's point of view. Therefore, sound source directivity is not discussed, even though it is an important topic on its own—e.g., it is essential to model it correctly in a six-degrees-of-freedom (6DoF) application where the listener is allowed to walk past a directional source [51].

2.1 The binaural case

When rendering reverberation binaurally, directional information of reflected sounds is encoded in the binaural room impulse response (BRIR), i.e., a pair of RIRs that are measured at the listener's ear canals, in the form of monaural and interaural cues. Therefore, the most effective and straightforward way to achieve an accurate binaural rendering is to convolve an anechoic audio signal with a BRIR. Static (non-head-tracked) BRIR-based renderings can produce highly authentic binaural signals, to the point of being indistinguishable from those emitted by real sound sources [52–55]. On the other hand, dynamic (head-tracked) renderings are more challenging to implement, as they require swapping between BRIRs as the listener or the source move. It is worth noting that, when dealing with binaural renderings of anechoic environments, an angular movement of a source relative to the listener is roughly equivalent to a head rotation of the listener, which is typically trivial to compute in the Ambisonics domain using rotation matrices ([56], Section 5.2.2). However, this does not generalise to reverberant environments, where the room

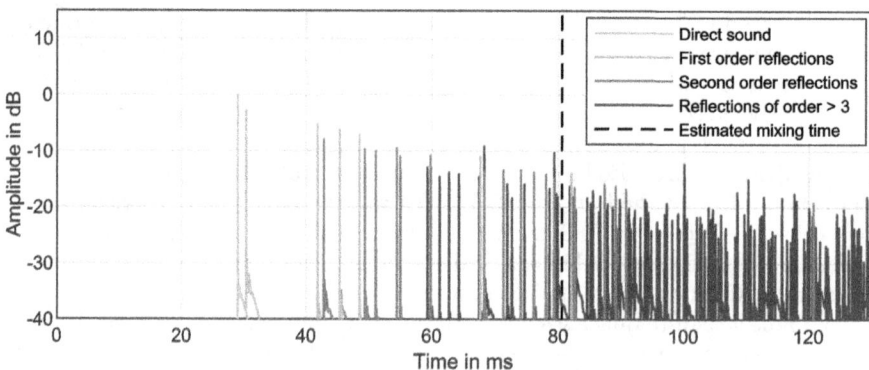

Figure 1.

First 130 ms of an RIR, expressed in decibels relative to the peak value. The RIR was simulated with the image source method [31] for an omnidirectional point source placed 10 m away from the receiver in a room with an approximate volume of 2342.7 m^3. The mixing time, estimated according to Lindau et al. [48], is indicated.

provides a frame of reference, and the angular movement of a source is not equivalent to rotating the listener's head.

A recent study has suggested that BRIRs should be measured by varying the listener position in increments of 5 cm or less in a three-dimensional grid (which can be a costly process) to achieve a dynamic convolution-based rendering in which the swapping is seamless to the listener [57]. Alternatively, one may start from a coarser spatial grid and interpolate BRIRs at intermediate positions. Unfortunately, BRIR interpolation is not trivial because the time and direction of arrival of each reflection may vary depending on the receiver's position, changing the BRIR's temporal structure across the grid. Nevertheless, recent studies have shown promising progress by employing dual-band approaches and heuristics to match early reflections in the time domain [58, 59]. On a related note, another active research topic is the extrapolation of RIRs in the Ambisonics domain for 6DoF applications (e.g., [60–63]), which is further discussed in Section 4.

Although BRIRs are mainly obtained through binaural measurements made on a person's or a mannequin's head [55], they may also be generated from RIRs that were either measured with microphone arrays [64–68] or simulated [28, 35]. This approach typically involves identifying individual reflections and their direction of arrival, e.g., with the help of the spatial decomposition method (SDM) [65], and then convolving each reflection with a head-related impulse response (HRIR) for the corresponding direction [69]—which is equivalent to a multiplication with a head-related transfer function (HRTF) in the frequency domain. However, rendering the full length of the BRIR this way can easily become expensive, which is why simplified models such as the aforementioned 'hybrid' one become important: we can just render a few early reflections accurately while modelling late reverberation as a stochastic, non-directional process, and still produce binaural signals that are not perceptually different from properly rendered ones. This has been recently shown by Brinkmann et al. [47], who suggested that accurately rendering just six early reflections plus stochastic late reverberation may be enough to produce auralisations that are perceptually indistinguishable from a fully-rendered reference, for a simulation of a shoebox-type room.

It should be noted that modelling late reverberation as isotropic is computationally inexpensive but may lead to noticeable degradation when simulating asymmetrical rooms (e.g., a long and narrow corridor) where late reverberation is highly directional [12]. For such cases, Alary et al. have proposed to employ directional feedback delay networks (DFDN) [39], which extend the functionality of traditional FDNs to spatial audio and allow to inexpensively produce non-uniform reverberation, so that the RT is direction-dependent. A downside of DFDNs is their inability to correctly reproduce early reflections, which should be modelled separately for best results.

Another simplification consists in quantising the direction of arrival of reflections by 'snapping' them to the closest neighbour in a predefined grid. This method is explored by Amengual Garí et al. [69], who found that an RIR may be quantised to just 14 directions in a Lebedev grid [70] and still be used to render binaural signals through SDM without perceptual degradation when compared to the original. The scattering delay network method (SDN) is based on a similar premise, quantising the RIR to as many directions as first-order reflections, e.g., six for a cuboid room, while obtaining good results in perceptual evaluations [32]. The rationale of SDN is that early reflections are computed accurately, while later ones are approximated with higher error as time advances, which is a sensible approach from a perceptual point of view. However, it might lead to an inaccurate late reverberation tail, which is why combining SDN with an inexpensive method for late reverberation simulation (e.g., DFDN) might be a promising alternative.

On the other hand, rather than generating separate BRIRs for each rendered sound source, one may also 'encode' the sum of all of them into a single sound field, and then reproduce it binaurally, e.g., by means of a set of virtual loudspeakers. That way, only the virtual loudspeaker signals must be binaurally rendered, independently of the number of sources that form the sound field. This is a convenient simplification when many sources are rendered at once. As mentioned earlier, typical loudspeaker-based sound field reproduction methods include vector-based amplitude panning [49] and high-order Ambisonics [50, 56, 71]. The latter is by far the more popular method for binaural rendering, given its efficient simulation of head rotations ([56], Section 5.2.2) and manipulation of spatial resolution [72]. However, the Ambisonics processing may have perceivable effects on the binaural signals, which are still being investigated. Recent research on this topic is reviewed in Section 3.

3. Binaural Ambisonics-based reverberation and spatial resolution

The spherical harmonics framework (known as Ambisonics in the context of audio production) allows to express a sound field as a continuous function on a sphere around the listener. Ambisonics sound fields are typically generated from microphone array recordings [73] or plane-wave-based simulations. Alternatively, it is often convenient to measure or simulate an Ambisonics RIR that can be convolved with any anechoic audio signal to generate the sound field, e.g., as in [74]. Once encoded in the Ambisonics domain, a sound field can be mirrored, warped or rotated around the listener through inexpensive algebraic operations [56]. Additionally, it is possible to modify its spatial resolution, which allows to reduce computational costs in the rendering process in exchange for potential perceptual degradation [72, 75, 76].

When a sound field is encoded in the Ambisonics domain, its spatial resolution is defined by its inherent 'truncation order', which is an integer equal or greater than zero. Higher-order signals have a larger number of channels and allow to produce binaural renderings with finer spatial resolution and sound sources that are easier to localise, while lower-order signals are more lightweight (fewer channels) and produce renderings with lower resolution and 'blurry' sources (see **Figure** 2). This was shown by Avni et al. [77], who argued that truncating the order of an Ambisonics signal affected the perception of spaciousness and timbre in the resulting binaural signals. Later, Bernschütz [66] reported that, in perceptual evaluations, listeners could not generally detect differences in binaural signals rendered from Ambisonics sound fields of order 11 and above. Then, Ahrens and Andersson [74] showed that an order of 8 might be sufficient to simulate lateral sound sources that are indistinguishable from BRIR-based renderings, but slight differences were perceived up to order 29 for frontal sound sources.

It has also been shown that the relation between spatial order and perceived quality also depends on the 'decoding' method that is used to translate the Ambisonics sound field to a pair of binaural signals. For instance, the time-alignment method [78] and the magnitude least squares (MagLS) method [79] have both been shown to produce more accurate binaural signals at lower spatial orders than other approaches, such as the widely used virtual loudspeakers method [80]. In the case of MagLS, which focuses on minimising magnitude errors (disregarding phase) at high frequencies, Sun [81] showed that a conceptually similar method was able to produce binaural signals that were indistinguishable from a high-order reference at orders as low as 14.

Overall, previous studies have suggested that binaural signals can be accurately rendered from Ambisonics sound fields as long as the truncation order is high

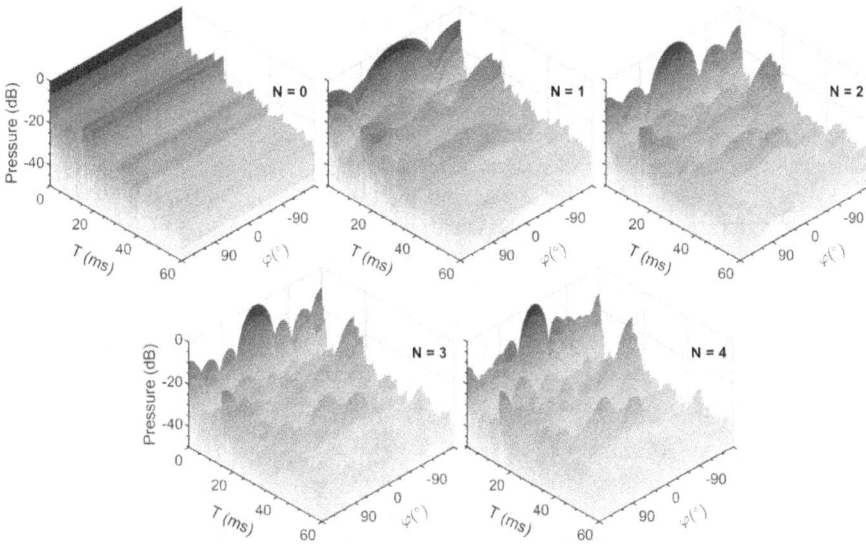

Figure 2.
Room impulse response encoded in the Ambisonics domain at different truncation orders (0 to 4), for a source placed in front of the listener. Data are plotted as sound pressure (in decibels relative to the peak value) along the time axis and over different azimuth angles on the horizontal plane. Source: Engel et al. [76] ('trapezoid' room).

enough, probably somewhere between 8 and 29. However, such orders may still be too high to be computationally efficient (the number of channels of an Ambisonics signal is proportional to the square of its truncation order) or just unfeasible in practice (commercially available microphone arrays operate at order 4 or lower). The remainder of this section discusses some recent perceptual studies that explored how the binaural rendering of reverberant sound fields is affected when simplifications are applied in the Ambisonics domain, e.g., reducing the truncation order of different parts of the RIR.

3.1 Hybrid Ambisonics

A recent listening experiment by Lübeck et al. [75] showed that early reflections and late reverberation may be encoded in Ambisonics at a significantly lower order than the direct sound and still produce binaural signals that are indistinguishable from a BRIR-based rendering. The reason why this may happen is illustrated in **Figure 2**, which shows an RIR encoded in Ambisonics at different truncation orders. It can be seen how the lowest order (0) produces an isotropic signal which does not vary across directions in the horizontal plane, while higher orders achieve a more faithful representation of the sound field by allowing for spatially 'sharper' patterns— e.g., note how the direct sound becomes narrower as order increases, converging towards a spatial Dirac delta. Looking at this figure, it becomes apparent that earlier parts of the RIR (blue) are more sensible to spatial resolution changes due to order truncation, compared to late reverberation (green) which is less directional.

According to these observations, it is reasonable to propose an Ambisonics-based binaural rendering method that employs a high truncation order for the direct sound (and, possibly, some early reflections) and lower orders for the rest of the RIR. Such a method could be highly efficient given that late reverberation usually accounts for the majority of the duration of the RIR. This approach, reminiscent to the hybrid models discussed earlier, has been tentatively coined as 'hybrid Ambisonics'.

A perceptual study by Engel et al. [76] evaluated binaural signals generated with hybrid Ambisonics and the virtual loudspeaker method, and found that an order between 2 or 3 (dependent on the room) may be enough to render reverberation, assuming that the direct sound path is accurately reproduced through convolution with HRIRs (see **Figure 3**). This is a promising precedent for future efficient binaural rendering methods, although further investigations would be needed to generalise these results to a wider selection of rooms and stimuli types. In the future, a more general model could estimate the needed truncation order adaptively based on the Ambisonics signal (e.g., measuring its directivity over time), which could be used in efficient binaural renderers or as a way to compress spatial audio data.

3.2 Reverberant virtual loudspeaker (RVL)

In real-time interactive binaural simulations, RIRs are typically recomputed when there is a change in the scene such as movements of the listener or sources. When working in the Ambisonics domain, this recomputation is not needed in order to simulate a head rotation from the listener, as the signal can be efficiently rotated via a rotation matrix ([56], Section 5.2.2). However, translational movements of either the listener or a source still require to recompute the RIRs. As a result, the number of sources that can be rendered simultaneously in a low-cost scenario might be limited.

In such cases, it may be beneficial to employ a rendering method that scales well with the number of sources. One such example is the reverberant virtual loudspeaker method (RVL), an Ambisonics-based approach that has the advantage of requiring a fixed amount of real-time convolutions regardless of the number of sources [72, 76, 83]. This method takes inspiration from the virtual loudspeakers approach [71, 80], which decodes an Ambisonics sound field to a virtual loudspeaker grid around the listener and convolves the resulting signals with HRIRs to generate the binaural output. RVL performs this same process but, instead of HRIRs, the virtual loudspeaker signals are convolved with BRIRs, so the acoustics of the room are effectively integrated with the binaural rendering without the need for additional steps. Therefore, the number of real-time convolutions depends only on the truncation order of the sound field, independently of the number of rendered

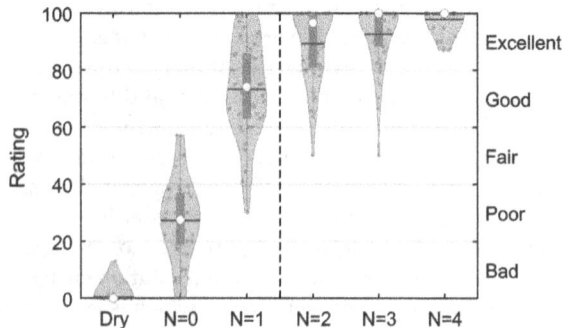

Figure 3.
*Perceptual ratings of binaural renderings generated from the hybrid-Ambisonics RIRs of orders 0 to 4 are shown in **Figure 2**, where the direct sound was reproduced via convolution with a single HRIR. A dry rendering was used as the anchor signal and the 4th order signal, as the reference. The vertical dotted lines indicate that the groups on the left are significantly different ($p < 0.05$) from the groups on the right. Source: Engel et al. [76].*

sources. For this reason, RVL is highly efficient at rendering a large number of sources in real time (see **Figure 4**). Its main limitation is that the room is head-locked due to the set of BRIRs being fixed, so head rotations may lead to inaccurate reflections, as shown in **Figure 5**.

RVL was perceptually evaluated in [76], paying particular attention to its effect on head rotations. For the assessment, the method was applied only to the rever-berant sound (direct sound was generated through convolution with HRIRs) and the implementation was done with the 3D Tune-In Toolkit spatial audio library [84]. Listeners were asked to compare RVL to first-order hybrid Ambisonics ren-derings (both head-tracked) of speech and music, by being asked 'Considering the given room [shown in a picture], which example is more appropriate?'. Results suggested that the inaccurate head rotations could indeed be detected by listeners but were not necessarily perceived as a degradation in quality with respect to the more accurate rendering—note the bimodal distribution shown in **Figure 6**, which indicates that there was not a unanimous preference towards either rendering.

One could speculate that the RVL method was preferred by some listeners due to the BRIR-based rendering leading to highly uncorrelated binaural signals, which are typically associated with higher perceived quality when evaluating late reverbera-tion (see the binaural quality index by Beranek [9]). An additional investigation to explore the matter further would be to compare the RVL method to other approaches that specifically aim to optimise interaural coherence, such as the

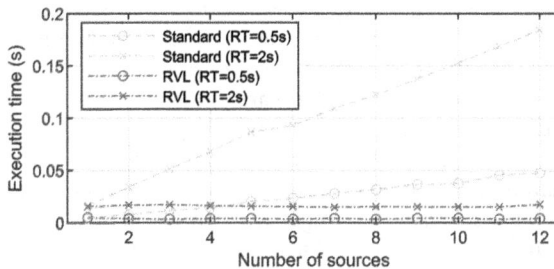

Figure 4.
Comparison between the average execution time of the convolution stage in Ambisonics binaural rendering ('standard') and RVL binaural rendering, as a function of the number of rendered sources, for two different reverberation times (RT). A random input signal with a length of 1024 samples was used as input. Simulations were done in MATLAB (MathWorks) using the overlap-add method [82], running on a quad-core processor at 2.8 GHz. Source: Engel et al. [76].

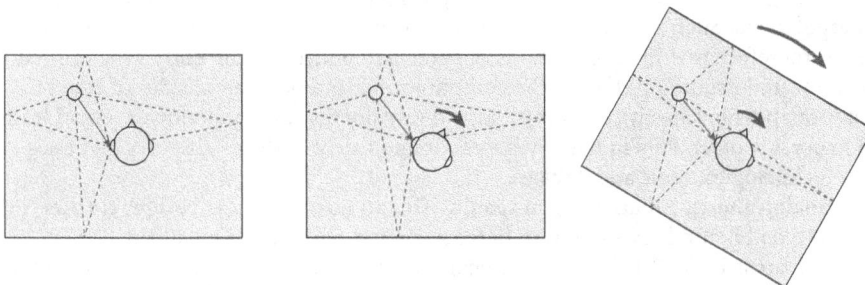

Figure 5.
Direct sound path and first-order early reflections as they reach the left ear of a listener in three scenarios: (left) before any head rotation; (middle) canonical rendering after a head rotation of 30 degrees clockwise; and (right) RVL rendering after the same head rotation. Note how, in the third scenario, the direct sound path is accurate, whereas the room is head-locked, affecting the incoming direction of reflections. Source: Engel et al. [76].

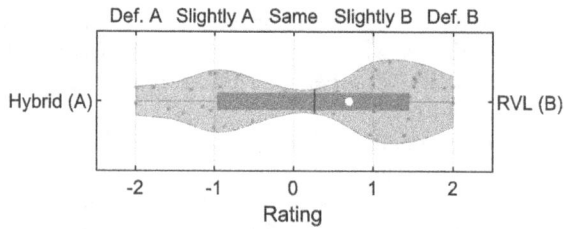

Figure 6.
Violin plot showing perceptual ratings from paired comparisons between first-order hybrid Ambisonics (A) and RVL (B) binaural renderings. Negative values represent preference towards a, while positive values represent preference towards B. Source: Engel et al. [76].

covariance constraint method proposed by Zaunschirm et al. [78] and described by Zotter and Frank ([56], Section 4.11.3).

Regardless, further perceptual evaluations (e.g., in more rooms) would be needed to generalise these results. Overall, RVL could be a viable option to render binaural reverberation of a large number of sources in real time in a low-resource scenario.

4. Future directions

The trade-off between complexity and perceived quality when rendering binaural reverberation is still an area of major interest that has to be further explored. Recent studies have looked at the perceptual impact of varying spatial resolution of Ambisonics-based reverberation, but there are yet aspects of it that warrant further research. For instance, it would be interesting to explore an approach to compress Ambisonics RIRs by truncating their order depending on their directional and temporal information, as a way to compute and store them more efficiently.

Another set of very relevant challenges will come from using artificial binaural reverberation in different contexts and tasks. For example, binaural audio has been used in the past for assisting blind individuals in learning about the spatial configuration of a closed environment before being physically introduced to it [85, 86]. Within that context, the creation of geometrically and spatially accurate real-time reverberation was extremely important and could be achieved only by performing a series of case-specific optimisations in the processing chain, for example, limiting navigation paths to a series of lines rather than a 2-dimensional space, and pre-calculating a set of Ambisonics RIRs computing in real-time only rotations and interpolations. Such optimisations can be allowed only within a research environment, therefore real-life applications of such techniques are currently very limited. A better understanding of both the computational and perceptual sides of reverberation, possibly specifically for blind and visually impaired individuals, could lead to major advancements in the development and use of auditory displays and assistive technologies, tools and devices.

Looking ahead, AR applications could offer an interesting testbed for further research on binaural reverberation perception and rendering. One of the key research areas in AR/VR is 6DoF (or position-dynamic) audio rendering, where the listener is allowed to move around the scene, as opposed to traditional Ambisonics rendering where only head rotations are allowed (three degrees of freedom). Several methods have been recently proposed to efficiently extrapolate spatial audio signals from one listener position to another, either via simple parametric methods

[87] or more complex Ambisonics-based approaches that often rely on parametrising the sound field in 'direct' and 'ambient' components [60, 61], or according to the source distance [62, 63]. Significant advancements have also been made in terms of recording complex auditory scenes and to make them navigable in 6DoF—in this case, specialised hardware and software has been released and is already available commercially [88]. Future improvements in 6DoF recording and rendering techniques will in turn allow for an increased level of interactivity within the simulation, as well as more effective evaluations of different audio rendering technologies using AR/VR systems.

Focusing on the AR case, in order to blend real with virtual audio, it is essential to develop techniques for the automatic estimation of the reverberant characteristics of the real environment. New methods will need to be developed and evaluated for blending virtual audio sources within real scenes and to evaluate the impact of blending accuracy through metrics related to perceived realism and scene acceptability. This can be achieved, for example, by characterising the acoustical environment surrounding the AR user, using this in-situ data to synthesise virtual sounds with matching acoustic properties. Machine learning (ML) techniques could be employed to address the issue of blind acoustical environment characterisation by focusing first on overall room fingerprint evaluation (late reverberation), then on the finer details of the room response that vary depending on specific source positions (early reflections). The scene analysis could also be used to extract the direction-of-arrival for multiple sound sources and direct-to-reverberant energy ratio by separating source information from room and user acoustic properties. The data extracted by the model could then be employed to generate realistic virtual reverberation, which will be matched with the real-world reverberation. Of course for each step of this scenario several open challenges still exist, both from the computational point of view (e.g., how to generate geometrically and directionally accurate reverberation in real-time) and from the perceptual point of view (e.g., what is perceptually relevant and should therefore be computationally modelled and rendered, and what can be approximated).

Better understanding the extent and origin of sensory thresholds in terms of reverberation perception, therefore, presents still a very open set of challenges, which will need to be addressed in the future through extensive listening experiments and, why not, also by means of binaural auditory models and ML-trained 'artificial listeners'.

5. Conclusions

Within this chapter, an overview was presented on perception and efficient simulation of reverberation. A special focus has been put on the case of binaural audio and, in particular, on Ambisonics-based and convolution-based rendering methods. The issue of the trade-off between computational cost and perceived quality has been discussed at length, mainly looking at the case of varying spatial resolution and implementation choices of Ambisonics-based renderings, highlighting the results of some recent studies on this matter. Considering the very rapid development and uptake of VR and AR technologies, it is particularly evident the importance of further research focusing on better understanding how computational optimisations and simplifications can have an impact on the perceived quality and realism of the rendering. Some of the most relevant challenges in this area have been outlined at the end of the chapter, and will hopefully serve as a guideline for future research in the area.

Acknowledgements

The writing of this chapter has been supported by the SONICOM project, which has received funding from the European Union's Horizon 2020 research and innovation programme under grant agreement no. 101017743.

Nomenclature

6DoF	Six degrees of freedom
AR	Augmented reality
BRIR	Binaural room impulse response
DFDN	Directional feedback delay network
FDN	Feedback delay network
HRIR	Head-related impulse response
HRTF	Head-related transfer function
IC	Interaural coherence
ILD	Interaural level difference
ML	Machine learning
RIR	Room impulse response
RT	Reverberation time
RVL	Reverberant virtual loudspeaker
SDM	Spatial decomposition method
SDN	Scattering delay network
VR	Virtual reality

Author details

Isaac Engel* and Lorenzo Picinali
Dyson School of Design Engineering, Imperial College London, London, United Kingdom

*Address all correspondence to: isaac.engel@imperial.ac.uk

IntechOpen

References

[1] Wallach H, Newman EB, Rosenzweig MR. A precedence effect in sound localization. The Journal of the Acoustical Society of America. 1949; 21(4):468-468. DOI: 10.1121/1.1917119

[2] Litovsky RY, Steven Colburn H, Yost WA, Guzman SJ. The precedence effect. The Journal of the Acoustical Society of America. 1999;106(4): 1633-1654. DOI: 10.1121/1.427914

[3] Brown AD, Christopher Stecker G, Tollin DJ. The precedence effect in sound localization. JARO: Journal of the Association for Research in Otolaryngology. 2015;16(1):1-28. DOI: 10.1007/s10162-014-0496-2

[4] Moore BCJ. An Introduction to the Psychology of Hearing. Leiden, Netherlands: Brill; 2012

[5] Olive SE, Toole FE. The detection of reflections in typical rooms. Journal of the Audio Engineering Society. 1989;37 (7/8):539-553. https://aes.org/e-lib/b rowse.cfm?elib=6079

[6] Bech S. Timbral aspects of reproduced sound in small rooms. II. The Journal of the Acoustical Society of America. 1996;99(6):3539-3549. DOI: 10.1121/1.414952

[7] Barron M, Marshall AH. Spatial impression due to early lateral reflections in concert halls: The derivation of a physical measure. Journal of Sound and Vibration. 1981; 77(2):211-232. DOI: 10.1016/ S0022-460X(81)80020-X

[8] Kaplanis N, Bech S, Jensen SH, van Waterschoot T. Perception of reverberation in small rooms: A literature study. In: Audio Engineering Society Conference: 55th International Conference: Spatial Audio. Helsinki, Finland: Audio Engineering Society; 2014.

https://aes.org/e-lib/browse.cfm?elib= 17348

[9] Beranek LL. Concert hall acoustics— 2008. Journal of the Audio Engineering Society. 2008;56(7/8):532-544. https:// aes.org/e-lib/browse.cfm?elib=14398

[10] International Organization for Standardization. Measurement of Room Acoustic Parameters — Part 1: Performance Spaces (ISO Standard No. 3382-1:2009). 2009. https://www.iso. org/standard/40979.html

[11] Yadav M, Cabrera DA, Miranda L, Martens WL, Lee D, Collins R. Investigating auditory room size perception with autophonic stimuli. In: Audio Engineering Society Convention 135. New York, USA: Audio Engineering Society; 2013. https://aes. org/e-lib/browse.cfm?elib=16984

[12] Alary B, Massé P, Välimäki V, Noisternig M. Assessing the anisotropic features of spatial impulse responses. In: EAA Spatial Audio Signal Processing Symposium. Paris, France: Sorbonne Université; 2019. pp. 43-48. DOI: 10.25836/sasp.2019.32

[13] Begault DR, Wenzel EM, Anderson MR. Direct comparison of the impact of head tracking, reverberation, and individualized head-related transfer functions on the spatial perception of a virtual speech source. Journal of the Audio Engineering Society. 2001; 49(10):904-916. http://aes.org/e-lib/b rowse.cfm?elib=10175

[14] Leclère T, Lavandier M, Perrin F. On the externalization of sound sources with headphones without reference to a real source. The Journal of the Acoustical Society of America. 2019;146(4): 2309-2320. DOI: 10.1121/1.5128325

[15] Catic J, Santurette S, Dau T. The role of reverberation-related binaural cues in

the externalization of speech. The Journal of the Acoustical Society of America. 2015;**138**(2):1154-1167. DOI: 10.1121/1.4928132

[16] Hassager HG, Gran F, Dau T. The role of spectral detail in the binaural transfer function on perceived externalization in a reverberant environment. The Journal of the Acoustical Society of America. 2016; **139**(5):2992-3000. DOI: 10.1121/1.4950847

[17] Jiang Z, Sang J, Zheng C, Li X. The effect of pinna filtering in binaural transfer functions on externalization in a reverberant environment. Applied Acoustics. 2020;**164**:107257. DOI: 10.1016/j.apacoust.2020.107257

[18] Engel I, Alon DL, Robinson PW, Mehra R. The effect of generic headphone compensation on binaural renderings. In: Audio Engineering Society Conference: 2019 AES International Conference on Immersive and Interactive Audio. York, UK: Audio Engineering Society; 2019. https://www. aes.org/e-lib/browse.cfm?elib=20387

[19] Engel I, Alon DL, Scheumann K, Mehra R. Listener-preferred headphone frequency response for stereo and spatial audio content. In: Audio Engineering Society Conference: 2020 AES International Conference on Audio for Virtual and Augmented Reality. Virtual Reality: Audio Engineering Society; 2020. https://aes.org/e-lib/browse.cfm?elib= 20868

[20] Catic J, Santurette S, Buchholz JM, Gran F, Dau T. The effect of interaural-level-difference fluctuations on the externalization of sound. The Journal of the Acoustical Society of America. 2013; **134**(2):1232-1241. DOI: 10.1121/1.4812264

[21] Li S, Schlieper R, Peissig J. The effect of variation of reverberation parameters in contralateral versus ipsilateral ear signals on perceived externalization of a lateral sound source in a listening room. The Journal of the Acoustical Society of America. 2018; **144**(2):966-980. DOI: 10.1121/1.5051632

[22] Li S, Schlieper R, Peissig J. The role of reverberation and magnitude spectra of direct parts in contralateral and ipsilateral ear signals on perceived externalization. Applied Sciences. 2019; **9**(3):460. DOI: 10.3390/app9030460

[23] Werner S, Klein F, Mayenfels T, Brandenburg K. A summary on acoustic room divergence and its effect on externalization of auditory events. In: 2016 Eighth International Conference on Quality of Multimedia Experience (QoMEX). Lisbon, Portugal: IEEE; 2016. pp. 1-6. DOI: 10.1109/QoMEX.2016. 7498973

[24] Werner S, Götz G, Klein F. Influence of head tracking on the externalization of auditory events at divergence between synthesized and listening room using a binaural headphone system. In: Audio Engineering Society Convention 142. Berlin, Germany: Audio Engineering Society; 2017. https://www.aes.org/ e-lib/browse.cfm?elib=18568

[25] Klein F, Werner S, Mayenfels T. Influences of training on externalization of binaural synthesis in situations of room divergence. Journal of the Audio Engineering Society. 2017;**65**(3): 178-187. https://www.aes.org/e-lib/b rowse.cfm?elib=18553

[26] Engel I, Goodman DFM, Picinali L. The effect of auditory anchors on sound localization: A preliminary study. In: Audio Engineering Society Conference: 2019 AES International Conference on Immersive and Interactive Audio. York, UK: Audio Engineering Society; 2019. https://aes.org/e-lib/browse.cfm?elib= 20388

[27] Best V, Baumgartner R, Lavandier M, Majdak P, Kopčo N. Sound externalization: A review of recent research. Trends in Hearing. 2020;**24**:2331216520948390. DOI: 10.1177/2331216520948390

[28] Schissler C, Stirling P, Mehra R. Efficient construction of the spatial room impulse response. In: 2017 IEEE Virtual Reality (VR). Los Angeles, USA: IEEE; 2017. pp. 122-130. DOI: 10.1109/VR.2017.7892239

[29] Farina A. Advancements in impulse response measurements by sine sweeps. In: Audio Engineering Society Convention 122. Vienna, Austria: Audio Engineering Society; 2007. https://aes.org/e-lib/browse.cfm?elib=14106

[30] Botteldooren D. Finite-difference time-domain simulation of low-frequency room acoustic problems. The Journal of the Acoustical Society of America. 1995;**98**(6):3302-3308. DOI: 10.1121/1.413817

[31] Allen JB, Berkley DA. Image method for efficiently simulating small-room acoustics. The Journal of the Acoustical Society of America. 1979;**65**(4):943-950. DOI: 10.1121/1.382599

[32] De Sena E, Hacıhabiboğlu H, Cvetković Z, Smith JO. Efficient synthesis of room acoustics via scattering delay networks. IEEE/ACM Transactions on Audio, Speech, and Language Processing. 2015;**23**(9): 1478-1492. DOI: 10.1109/TASLP.2015.2438547

[33] Vorländer M. Simulation of the transient and steady-state sound propagation in rooms using a new combined ray-tracing/image-source algorithm. The Journal of the Acoustical Society of America. 1989;**86**(1):172-178. DOI: 10.1121/1.398336

[34] Lentz T, Schröder D, Vorländer M, Assenmacher I. Virtual reality system with integrated sound field simulation and reproduction. EURASIP Journal on Advances in Signal Processing. 2007; **2007**(1):187. DOI: 10.1155/2007/70540

[35] Schissler C, Mehra R, Manocha D. High-order diffraction and diffuse reflections for interactive sound propagation in large environments. ACM Transactions on Graphics. 2014;**33** (4):1-12. DOI: 10.1145/2601097.2601216

[36] Schroeder MR, Logan BF. "Colorless" artificial reverberation. IRE Transactions on Audio, AU. 1961;**9**(6):209-214. DOI: 10.1109/TAU.1961.1166351

[37] Jot J-M, Chaigne A. Digital delay networks for designing artificial reverberators. In: Audio Engineering Society Convention 90. Paris, France: Audio Engineering Society; 1991. https://aes.org/e-lib/browse.cfm?elib=5663

[38] Jot J-M. Efficient models for reverberation and distance rendering in computer music and virtual audio reality. In: ICMC: International Computer Music Conference. Ann Arbor, Michigan, USA: Thessaloniki, Greece; 1997. pp. 236-243. https://hal.archives-ouvertes.fr/hal-01106168

[39] Alary B, Politis A, Schlecht S, Välimäki V. Directional feedback delay network. Journal of the Audio Engineering Society. 2019;**67**(10): 752-762. http://www.aes.org/e-lib/browse.cfm?elib=20693

[40] Murphy DT, Stewart R. A hybrid artificial reverberation algorithm. In: Audio Engineering Society Convention 122. Vienna, Austria: Audio Engineering Society; 2007. https://aes.org/e-lib/browse.cfm?elib=14006

[41] Carpentier T, Noisternig M, Warusfel O. Hybrid reverberation processor with perceptual control. In: 17th International Conference on Digital Audio Effects - DAFx-14. Erlangen,

Germany: International Audio Laboratories Erlangen; 2014. pp. 93-100. https://hal.archives-ouvertes.fr/hal-01107075

[42] Coleman P, Franck A, Jackson PJB, Hughes RJ, Remaggi L, Melchior F. Object-based reverberation for spatial audio. Journal of the Audio Engineering Society. 2017;**65**(1/2):66-77. https://www.aes.org/e-lib/browse.cfm?elib=18544

[43] Coleman P, Franck A, Menzies D, Jackson PJB. Object-based reverberation encoding from first-order ambisonic RIRs. In: Audio Engineering Society Convention 142. Boston, Massachusetts, USA: Audio Engineering Society; 2017. https://www.aes.org/e-lib/browse.cfm?elib=18608

[44] Stade P, Arend JM, Pörschmann C. A parametric model for the synthesis of binaural room impulse responses. In: Proceedings of Meetings on Acoustics. Vol. 30. Boston, Massachusetts, USA: Acoustical Society of America; 2017. p. 015006. DOI: 10.1121/2.0000573

[45] Raghuvanshi N, Snyder J. Parametric directional coding for precomputed sound propagation. ACM Transactions on Graphics. 2018;**37**(4): 1-14. DOI: 10.1145/3197517.3201339

[46] Godin K, Gamper H, Raghuvanshi N. Aesthetic modification of room impulse responses for interactive auralization. In: Audio Engineering Society Conference: 2019 AES International Conference on Immersive and Interactive Audio. York, UK: Audio Engineering Society; 2019. https://www.aes.org/e-lib/browse.cfm?elib=20444

[47] Brinkmann F, Gamper H, Raghuvanshi N, Tashev I. Towards encoding perceptually salient early reflections for parametric spatial audio rendering. In: Audio Engineering Society Convention 148. Audio

Engineering Society; 2020. https://aes.org/e-lib/browse.cfm?elib=20797

[48] Lindau A, Kosanke L, Weinzierl S. Perceptual evaluation of model- and signal-based predictors of the mixing time in binaural room impulse responses. Journal of the Audio Engineering Society. 2012;**60**(11): 887-898. https://aes.org/e-lib/browse.cfm?elib=16633

[49] Pulkki V. Virtual sound source positioning using vector base amplitude panning. Journal of the Audio Engineering Society. 1997;**45**(6): 456-466. https://www.aes.org/e-lib/browse.cfm?elib=7853

[50] Gerzon MA. Periphony: With-height sound reproduction. Journal of the Audio Engineering Society. 1973; **21**(1):2-10. https://www.aes.org/e-lib/browse.cfm?elib=2012

[51] Werner S, Klein F, Neidhardt A, Sloma U, Schneiderwind C, Brandenburg K. Creation of auditory augmented reality using a position-dynamic binaural synthesis system— Technical components, psychoacoustic needs, and perceptual evaluation. Applied Sciences. 2021;**11**(3):1150. DOI: 10.3390/app11031150

[52] Langendijk EHA, Bronkhorst AW. Fidelity of three-dimensional-sound reproduction using a virtual auditory display. The Journal of the Acoustical Society of America. 1999;**107**(1): 528-537. DOI: 10.1121/1.428321

[53] Moore AH, Tew AI, Nicol R. An initial validation of individualized crosstalk cancellation filters for binaural perceptual experiments. AES: Journal of the Audio Engineering Society. 2010;**58** (1–2):36-45

[54] Masiero B, Fels J. Perceptually robust headphone equalization for binaural reproduction. In: Audio Engineering Society Convention 130.

London, UK: Audio Engineering Society; 2011. pp. 1-7. DOI: 10.13140/2.1.1598.6882

[55] Brinkmann F, Lindau A, Weinzierl S. On the authenticity of individual dynamic binaural synthesis. The Journal of the Acoustical Society of America. 2017;**142**(4):1784-1795. DOI: 10.1121/1.5005606

[56] Zotter F, Frank M. Ambisonics: A Practical 3D Audio Theory for Recording, Studio Production, Sound Reinforcement, and Virtual Reality. Basingstoke, UK: Springer Nature; 2019. DOI: 10.1007/978-3-030-17207-7

[57] Neidhardt A, Reif B. Minimum BRIR grid resolution for interactive position changes in dynamic binaural synthesis. In: Audio Engineering Society Convention 148. Audio Engineering Society; 2020. https://www.aes.org/e-lib/browse.cfm?elib=20788

[58] Zaunschirm M, Zotter F, Frank M. Perceptual evaluation of variable-orientation binaural room impulse response rendering. In: Audio Engineering Society Conference: 2019 AES International Conference on Immersive and Interactive Audio. York, UK: Audio Engineering Society; 2019. https://www.aes.org/e-lib/browse.cfm?elib=20395

[59] Bruschi V, Nobili S, Cecchi S, Piazza F. An innovative method for binaural room impulse responses interpolation. In: Audio Engineering Society Convention 148. Audio Engineering Society; 2020. https://www.aes.org/e-lib/browse.cfm?elib=20802

[60] Kentgens M, Behler A, Jax P. Translation of a higher order ambisonics sound scene based on parametric decomposition. In: ICASSP 2020–2020 IEEE International Conference on Acoustics, Speech and Signal Processing (ICASSP). Barcelona, Spain: IEEE; 2020.

pp. 151-155. DOI: 10.1109/ICASSP40776.2020.9054414

[61] Müller K, Zotter F. Auralization based on multi-perspective ambisonic room impulse responses. Acta Acustica. 2020;4(6):25. DOI: 10.1051/aacus/2020024

[62] Kentgens M, Jax P. Translation of a higher-order ambisonics sound scene by space warping. In: Audio Engineering Society Conference: 2020 AES International Conference on Audio for Virtual and Augmented Reality. Virtual Reality: Audio Engineering Society; 2020. https://www.aes.org/e-lib/browse.cfm?elib=20873

[63] Birnie L, Abhayapala T, Tourbabin V, Samarasinghe P. Mixed source sound field translation for virtual binaural application with perceptual validation. IEEE/ACM Transactions on Audio, Speech, and Language Processing. 2021;**29**: 1188-1203. DOI: 10.1109/TASLP.2021.3061939

[64] Merimaa J, Pulkki V. Spatial impulse response rendering I: Analysis and synthesis. Journal of the Audio Engineering Society. 2005;**53**(12): 1115-1127. https://aes.org/e-lib/browse.cfm?elib=13401

[65] Tervo S, Pätynen J, Kuusinen A, Lokki T. Spatial decomposition method for room impulse responses. Journal of the Audio Engineering Society. 2013;**61** (1/2):17-28. http://www.aes.org/e-lib/browse.cfm?elib=16664

[66] Bernschütz B. Microphone Arrays and Sound Field Decomposition for Dynamic Binaural Recording. [Doctoral thesis]. Berlin, Germany: Technische Universität Berlin; 2016. DOI: 10.14279/depositonce-5082

[67] Zaunschirm M, Frank M, Zotter F. BRIR synthesis using first-order

microphone arrays. In: Audio Engineering Society Convention 144. Milan, Italy: Audio Engineering Society; 2018. https://www.aes.org/e-lib/browse.cfm?elib=19461

[68] Amengual Garí SV, Brimijoin WO, Hassager HG, Robinson PW. Flexible binaural resynthesis of room impulse responses for augmented reality research. In: EAA Spatial Audio Signal Processing Symposium. Paris, France: Sorbonne Université; 2019. pp. 161-166. DOI: 10.25836/sasp.2019.31

[69] Amengual Garí SV, Arend JM, Calamia PT, Robinson PW. Optimizations of the spatial decomposition method for binaural reproduction. Journal of the Audio Engineering Society. 2021;**68**(12): 959-976. https://www.aes.org/e-lib/browse.cfm?elib=21010

[70] Vyacheslav Ivanovich Lebedev. Spherical quadrature formulas exact to orders 25–29. Siberian Mathematical Journal, 18(1):99–107, 1977. https://https://link.springer.com/content/pdf/10.1007/BF00966954.pdf.

[71] Noisternig M, Musil T, Sontacchi A, Holdrich R. 3D binaural sound reproduction using a virtual ambisonic approach. In: IEEE International Symposium on Virtual Environments, Human-Computer Interfaces and Measurement Systems. Lugano, Switzerland: IEEE; 2003. VECIMS '03, 2003. pp. 174-178. DOI: 10.1109/VECIMS.2003.1227050

[72] Engel I, Henry C, Amengual Garí SV, Robinson PW, Poirier-Quinot D, Picinali L. Perceptual comparison of ambisonics-based reverberation methods in binaural listening. In: EAA Spatial Audio Signal Processing Symposium. Paris, France: Sorbonne Université; 2019. pp. 121-126. DOI: 10.25836/sasp.2019.11

[73] Rafaely B. Analysis and design of spherical microphone arrays. IEEE Transactions on Speech and Audio Processing. 2005;**13**(1):135-143. DOI: 10.1109/TSA.2004.839244

[74] Ahrens J, Andersson C. Perceptual evaluation of headphone auralization of rooms captured with spherical microphone arrays with respect to spaciousness and timbre. The Journal of the Acoustical Society of America. 2019; **145**(4):2783-2794. DOI: 10.1121/1.5096164

[75] Lübeck T, Pörschmann C, Arend JM. Perception of direct sound, early reflections, and reverberation in auralizations of sparsely measured binaural room impulse responses. In: Audio Engineering Society Conference: 2020 AES International Conference on Audio for Virtual and Augmented Reality. Virtual Reality: Audio Engineering Society; 2020. https://aes.org/e-lib/browse.cfm?elib=20865

[76] Engel I, Henry C, Amengual Garí SV, Robinson PW, Picinali L. Perceptual implications of different Ambisonics-based methods for binaural reverberation. The Journal of the Acoustical Society of America. 2021;**149**(2):895-910. DOI: 10.1121/10.0003437

[77] Avni A, Ahrens J, Geier M, Spors S, Wierstorf H, Rafaely B. Spatial perception of sound fields recorded by spherical microphone arrays with varying spatial resolution. The Journal of the Acoustical Society of America. 2013;**133**(5):2711-2721. DOI: 10.1121/1.4795780

[78] Zaunschirm M, Schörkhuber C, Höldrich R. Binaural rendering of Ambisonic signals by head-related impulse response time alignment and a diffuseness constraint. The Journal of the Acoustical Society of America. 2018; **143**(6):3616-3627. DOI: 10.1121/1.5040489

[79] Schörkhuber C, Zaunschirm M, Höldrich R. Binaural rendering of ambisonic signals via magnitude least squares. In: Fortschritte Der Akustik–DAGA. Munich, Germany: Deutsche Gesellschaft für Akustik; 2018. pp. 339-342. https://researchgate.net/publication/325080691_Binaural_Rendering_of_Ambisonic_Signals_via_Magnitude_Least_Squares

[80] McKeag A, McGrath DS. Sound Field Format to Binaural Decoder with Head Tracking. In: Audio Engineering Society Convention 6r. Melbourne, Australia: Audio Engineering Society; 1996. https://aes.org/e-lib/browse.cfm?elib=7477

[81] Sun D. Generation and Perception of Three-Dimensional Sound Fields Using Higher Order Ambisonics. [Doctoral thesis]. Sydney, Australia: University of Sydney; 2012

[82] Oppenheim AV, Buck JR, Schafer RW. Discrete-Time Signal Processing. Vol. 2. Upper Saddle River, N.J: Prentice Hall; 2001

[83] Picinali L, Wallin A, Levtov Y, Poirier-Quinot D. Comparative perceptual evaluation between different methods for implementing reverberation in a binaural context. In: Audio Engineering Society Convention 142. Berlin, Germany: Audio Engineering Society; 2017. https://hal.archives-ouvertes.fr/hal-01790217

[84] Cuevas-Rodríguez M, Picinali L, González-Toledo D, Garre C, de la Rubia-Cuestas E, Molina-Tanco L, et al. 3D Tune-In Toolkit: An open-source library for real-time binaural spatialisation. PLoS One. 2019;**14**(3): e0211899. DOI: 10.1371/journal.pone.0211899

[85] Katz BFG, Picinali L. Spatial audio applied to research with the blind. In:

Advances in Sound Localization. London, UK: InTech; 2011. pp. 225-250

[86] Picinali L, Afonso A, Denis M, Katz BFG. Exploration of architectural spaces by blind people using auditory virtual reality for the construction of spatial knowledge. International Journal of Human-Computer Studies. 2014;**72**(4): 393-407. DOI: 10.1016/j.ijhcs.2013.12.008

[87] Pörschmann C, Stade P, Arend JM. Binauralization of Omnidirectional Room Impulse Responses – Algorithm and Technical Evaluation. In: 20th International Conference on Digital Audio Effects. Edinburgh, UK: University of Edinburgh; 2017. http://dafx.de/paper-archive/2017/papers/DAFx17 paper 25.pdf

[88] Ciotucha T, Ruminski A, Zernicki T, Mróz B. Evaluation of six degrees of freedom 3d audio orchestra recording and playback using multi-point ambisonics interpolation. In: Audio Engineering Society Convention 150. Audio Engineering Society; 2021. https://www.aes.org/e-lib/browse.cfm?elib=21052

Section 3

Applications

Chapter 6

Binaural Reproduction Based on Bilateral Ambisonics

Zamir Ben-Hur, David Alon, Or Berebi, Ravish Mehra and Boaz Rafaely

Abstract

Binaural reproduction of high-quality spatial sound has gained considerable interest with the recent technology developments in virtual and augmented reality. The reproduction of binaural signals in the Spherical-Harmonics (SH) domain using Ambisonics is now a well-established methodology, with flexible binaural processing realized using SH representations of the sound-field and the Head-Related Transfer Function (HRTF). However, in most practical cases, the binaural reproduction is order-limited, which introduces truncation errors that have a detrimental effect on the perception of the reproduced signals, mainly due to the truncation of the HRTF. Recently, it has been shown that manipulating the HRTF phase component, by ear-alignment, significantly reduces its effective SH order while preserving its phase information, which may be beneficial for alleviating the above detrimental effect. Incorporating the ear-aligned HRTF into the binaural reproduction process has been suggested by using Bilateral Ambisonics, which is an Ambisonics representation of the sound-field formulated at the two ears. While this method imposes challenges on acquiring the sound-field, and specifically, on applying head-rotations, it leads to a significant reduction in errors caused by the limited-order reproduction, which yields a substantial improvement in the perceived binaural reproduction quality even with first order SH.

Keywords: binaural reproduction, HRTF, spherical-harmonics, 3D audio, spatial audio, ambisonics, head-tracking

1. Introduction

Recent developments in the field of virtual and augmented reality have increased the demand for high fidelity binaural reproduction technology [1]. Such technology aims to reproduce the spatial sound scene at the listener's ears through a pair of headphones, providing an immersive virtual sound experience [2]. The two main acoustic processes producing the binaural signals are the spatial sound-field result of the propagation from the sound source to the listener, and the interaction of this sound-field with the listener's body, which is described by the Head-Related Transfer Function (HRTF)[1] [3]. Binaural signals can be obtained directly using binaural microphones at the listener's ears [4]. In this way, the sound-field and the

---

[1] The term" HRTF" is used in this chapter to refer to the set of transfer functions for a set of source positions, unless stated otherwise.

HRTF are jointly captured and the reproduced binaural signals are limited to the recording scenario. More flexible reproduction, enabling, for example, the use of individual (personalized) HRTFs and head-tracking, can be obtained by rendering the binaural signals in post-processing. This requires the sound-field and the HRTF to be available separately. The HRTF could be obtained from an online database, or it could be measured acoustically or simulated numerically for an individual listener [5]. The sound field could also be simulated numerically, or captured using a microphone array [6–8].

In the past, the rendering of binaural signals using Ambisonics representation of the sound-field has been proposed [9–11]. The Ambisonics signals are the Spherical-Harmonics (SH) domain coefficients of the plane-wave amplitude density function, which encode the directional information of the sound-field. The binaural signals are computed by summing the products of the Ambisonics signals and the SH representation of the free-field HRTF. This offers the flexibility to manipulate either the sound field or the HRTF or both by employing algorithms that operate in the SH domain [12, 13].

The Ambisonics signals of a measured sound-field can be obtained from spherical microphone array recordings [14]. In practice, these arrays have a limited number of microphones, which limits the usable SH order [15]. A similar order limitation may also apply for a simulated sound-field due to computational efficiency considerations or memory usage [1, 16]. This order limitation places a constraint on the maximum SH order of the employed HRTF, which leads to truncation error [17]. Truncation error results in significant artifacts, both in frequency and in space, which have a detrimental effect on the perception of the reproduced binaural signals, for example, on the localization, source width, coloration and stability of the virtual sound source [18–21]. One way to overcome the limitations of low order Ambisonics is by a parametric representation of the sound field. For example, using DirAC [22], COMPASS [23], SPARTA [24] or HARPEX [25]. However, these approaches may introduce errors due to incomplete parameterization and thus do not provide ideal solution.

The HRTF truncation error can be reduced by pre-processing that lowers its effective SH order [26]. Evans et al. [27] suggested aligning the HRTF in the time domain prior to deriving its SH decomposition, and showed that this reduces the effective SH order significantly. They also showed that representing separately the magnitude and the unwrapped phase of the HRTF results in a lower SH order for both, compared to the complex-frequency representation. Romigh *et al.* [28] suggested using minimum-phase representation of the HRTF, together with logarithmic representation of the magnitude, and showed that a SH order as low as 4 is sufficient in order to achieve localization performance that is comparable with that of real sound sources in free-field. Brinkmann and Weinzierl [26] compared between these methods (among others), and concluded that the time-alignment method requires the lowest SH order in terms of SH energy distribution and Just Noticeable Difference (JND) in binaural models for source localization, coloration and correlation. Recently, a new method for efficient SH representation of HRTFs, which is based on ear-alignment, was presented [29]. This method proved to be more robust than the time-alignment method, while achieving a similar reduction in the effective SH order.

The order reduction of the HRTF using all the above methods is based on manipulating its phase component. However, the use of such a pre-processed HRTF for binaural reproduction using Ambisonics signals is not trivial due to the relation between the phases of the HRTF and the sound-field; hence, alternative solutions have also been explored. In [30], Zaunschirm et al. presented a method that uses a pre-processed HRTF, obtained by means of frequency-dependent time-alignment, to reproduce binaural signals in the SH domain using constrained optimization. They suggested pre-processing of the HRTF by removing its linear-phase component at high frequencies. Schörkhuber *et al.* further developed this approach in [31], where they presented the Magnitude Least-Squares (MagLS) method that performs magnitude-only optimization at high frequencies. Although the linear-phase component at high frequencies may be less important for lateral localization [32, 33], its removal still introduces errors in the binaural signal, and may affect other perceptual attributes [34, 35]. In [36], Lübeck *et al.* showed that the MagLS method achieved similar perceptual improvement to previously suggested diffuse field equalization methods for binaural reproduction [19, 37]. In [38], Jot *et al.* presented the Binaural B-Format approach, which uses first order Ambisonics signals at the location of the listener's ears and a minimum-phase approximation of the HRTF to compute the binaural signals directly at the listener's ears. This approach was further studied in [39, 40], along with several other approaches also based on the linear decomposition of the HRTF over spatial functions. Recently, the Binaural B-Format was extended to an arbitrary SH order using Bilateral Ambisonics reproduction [41, 42], which uses the ear-aligned HRTF and preserves the HRTF phase information. This method significantly reduces the truncation error and was shown to outperform current state-of-the-art methods using MagLS with low SH order reproduction. However, using Bilateral Ambisonics imposes challenges on acquiring the sound-field, and, specifically, on applying head-rotations to the reproduced binaural signal.

This chapter presents a detailed description of the Bilateral Ambisonics method, from HRTF representation to reproduction, including a possible solution for head tracking. The performance of the method is evaluated and compared with current state-of-the-art methods.

2. Basic ambisonics reproduction

This section provides an overview of the currently used formulation for binaural reproduction using Ambisonics signals, denoted here as Basic Ambisonics. The binaural signal, which is the sound pressure observed at each of the listener's ears, can be calculated, in the general case of a sound-field composed of a continuum of plane-waves, by [7, 16]:

$$p^{L\backslash R}(k) = \int_{\Omega \in S^2} a(k, \Omega) h^{L\backslash R}(k, \Omega) \, d\Omega, \tag{1}$$

where $a(k, \Omega)$ is the plane-wave amplitude density function, $\Omega \equiv (\theta, \phi) \in S^2$ is the spatial angle in standard spherical coordinates, with elevation angle $\theta \in [0, \pi]$, which is measured downwards from the Cartesian z axis, and azimuth angle $\phi \in [0, 2\pi)$, which is measured counter-clockwise from the Cartesian x axis in the xy-plane. $k = 2\pi f/c$ is the wave number, f is the frequency, and c is the speed of sound. $h^{L\backslash R}(k, \Omega)$ is the left ear, L, or right ear, R, HRTF, which is the acoustic transfer function from a far-field sound source to the listener's ear [3]. $p^{L\backslash R}(k)$ is the sound pressure at the ear and $\int_{\Omega \in S^2} (\cdot) \, d\Omega \equiv \int_0^{2\pi} \int_0^{\pi} (\cdot) \sin(\theta) \, d\theta \, d\phi$.

Alternatively, the binaural signal can be calculated in the SH domain, leading to the Basic Ambisonics reproduction formulation [10]:

$$p^{L\backslash R}(k) = \sum_{n=0}^{\infty} \sum_{m=-n}^{n} [\tilde{a}_{nm}(k)]^* h_{nm}^{L\backslash R}(k), \tag{2}$$

where $h_{nm}^{L\backslash R}(k)$ are the SH coefficients of the HRTF, which can be computed by applying the spherical Fourier transform (SFT) to the HRTF, $h^{L\backslash R}(k, \Omega)$. $\tilde{a}_{nm}(k)$ are the Basic Ambisonics signals, which are the SFT of $[a(k, \Omega)]^*$, where $[\cdot]^*$ denotes the complex conjugate. These Ambisonics signals can be calculated by capturing the sound-field using a spherical microphone array, and applying plane-wave decomposition in the SH domain [14, 43].

In practice, the infinite summation in Eq. (2) will be order limited:

$$p^{L\backslash R}(k) = \sum_{n=0}^{N} \sum_{m=-n}^{n} [\tilde{a}_{nm}(k)]^* h_{nm}^{L\backslash R}(k), \tag{3}$$

with $N = \min(N_a, N_h)$ [44], where N_a and N_h are the maximum available order of the Ambisonics signals and the HRTF, respectively. For example, when the Ambisonics signal is derived from spherical microphone array recordings, such as the Eigenmike [45], its order will be limited by the number of microphones; for the Eigenmike case with 32 microphones its order is around $N_a = 4$ [46]. A similar order limitation may also be introduced for a simulated sound-field in practical applications. On the other hand, Zhang et al. [17] showed that the HRTF is inherently of high spatial order. They concluded that for physically accurate representation up to 20 kHz, an order of above $N_h = 40$ is required. Therefore, in the practical scenario of $N_a = 4$, the HRTF will be severely truncated by the reproduction order, $N = 4$. This order truncation was shown to have a detrimental effect on the perceived spatial sound quality [18, 19], by affecting both the spectral and the spatial characteristics of the binaural signal.

3. Basic vs. ear-aligned HRTF representations

An efficient representation of the HRTF that reduces its effective SH order could provide a solution for reducing the effect of the truncation error on the reproduced binaural signal, caused by the limited order HRTF.

Recently, several pre-processing methods have been developed with the aim of reducing the effective SH order of the HRTF: for example, by time-alignment [27, 30], using directional equalization [47], using minimum-phase representation [28], or by ear-alignment [29, 48]. All these methods are based on manipulating the linear-phase component of the HRTF, which was shown to be the main contributor to the high-order nature of the HRTF [27].

Ear-alignment has been shown to be a robust method for reducing the effective SH order of the HRTF, while preserving the HRTF phase information and the Interaural Time Difference (ITD) [29], which are both important cues for sound source localization [5]. The alignment is performed by translating the origin of the free-field component of the HRTF from the center of the head to the position of the ear. This translation significantly reduces the effective SH order of the HRTF, as described next.

3.1 The effect of dual-centering on the basic SH representation of the HRTF

We denote the SH representation of the HRTF as the" basic representation". In this section, the effect of translating the origin of the free-field component of the HRTF on the basic representation is presented. This is performed by analyzing the simple case of a" free-field HRTF" as outlined in [29].

A pair of far-field HRTFs, h^L and h^R, is defined as a function of direction, Ω, and wave-number, k, by [3]:

$$h^{L\backslash R}(k, \Omega) = \frac{P^{L\backslash R}(k, \Omega)}{P_0(k, \Omega)},$$

(4)

where P^L and P^R represent the sound pressure at the left and right ears, respectively, and P_0 represents the free-field sound pressure at the center of the head in the absence of the head.

Now, consider a single plane-wave in free-field arriving from direction Ω with unit amplitude and wave number k. The sound pressure at position (Ω_0, r), can be written as [49]:

$$P_0(\Omega, k, \Omega_0, r) = e^{ikr\cos\Theta}$$

$$= \sum_{n=0}^{\infty} \sum_{m=-n}^{n} 4\pi i^n j_n(kr) \left[Y_n^m(\Omega)\right]^* Y_n^m(\Omega_0),$$

(5)

where Θ is the angle between Ω and Ω_0, $Y_n^m(\cdot)$ is the complex SH basis function of order n and degree m [50], and $j_n(\cdot)$ is the spherical Bessel function.

Defining the position of the ear to be at $(\Omega_{L\backslash R}, r_a)$, where r_a is the radius of the head, the free-field HRTF (an HRTF with the head absent) is defined by substituting Eq. (5) in Eq. (4):

$$h_0^{L\backslash R}(k, \Omega) = \frac{P_0\left(\Omega, k, \Omega_{L\backslash R}, r_a\right)}{P_0(\Omega, k, \Omega_0, 0)} = P_0\left(k, \Omega, \Omega_{L\backslash R}, r_a\right),$$

(6)

where $P_0(k, \Omega, \Omega_0, 0) = 1$ for all (Ω, k). Thus, for a sound-field composed of plane-waves from directions $\Omega \in \mathbb{S}^2$ the free-field HRTF can be written as:

$$h_0^{L\backslash R}(k, \Omega) = \sum_{n=0}^{\infty} \sum_{m=-n}^{n} 4\pi i^n j_n(kr_a) \left[Y_n^m(\Omega)\right]^* Y_n^m(\Omega_{L\backslash R}).$$

(7)

From here, the SH coefficients of the free-field HRTF can be derived, as presented in [29]:

$$h_{nm\,0}^{L\backslash R}(k) = 4\pi i^n j_n(kr_a) \left[Y_n^m(\Omega_{L\backslash R})\right]^*.$$

(8)

This equation provides insight into the potential effect of the dual-centering measurement process of the HRTF. The free-field HRTF coefficients, as described in the equation, have energy at every order n, which means that the HRTF is of infinite SH order. Nevertheless, it can be considered to be approximately order limited by $N_h = kr_a$, where $\cdot$ is the ceiling function, due to the behavior of the spherical Bessel function, which has a negligible magnitude for $kr >> n$ [49, 51]. On the other hand, from Eq. (6) it is clear that if the position of the ear was defined as the origin of the coordinate system, with $r_a = 0$, the free-field HRTF would be constant with unity value, which is represented by a zero order SH. This demonstrates how a sound pressure measured at a distance r_a from the origin, when normalized by a sound pressure at the origin, can lead to an increase in the SH order by approximately $N = \lceil kr_a \rceil$. An example of this added order is illustrated in **Figure 1**, which demonstrates how the SH order increases up to 30 at high frequencies.

Note the similarity of the orders in **Figure 1** to the actual order of the HRTFs as presented in [17], which suggests that although the explanation presented in this section is theoretical, it gives an insight into the possible increase in SH order due to the dual-centering of the HRTF definition.

3.2 HRTF ear-alignment

To compensate for the effect described in the previous section, with the aim of reducing the effective SH order of the HRTF, ear-alignment of the HRTF is suggested. The ear-aligned HRTF, h_a, is defined in a similar way to Eq. (4) as:

$$h_a^{L\backslash R}(k,\Omega) = \frac{P^{L\backslash R}(k,\Omega)}{P_0^{L\backslash R}(k,\Omega)}, \tag{9}$$

where $P_0^{L\backslash R}$ is the free-field pressure at the position of the left ear, L, or right ear, R, with the head absent. A measured HRTF can be aligned by translating the free-field pressure (denominator in (9)) from the center of the head to the position of the ear by:

$$h_a^{L\backslash R}(k,\Omega) = h^{L\backslash R}(k,\Omega) \cdot \frac{P_0(k,\Omega)}{P_0^{L\backslash R}(k,\Omega)}. \tag{10}$$

For a far-field HRTF, the free-field sound pressure can be computed using the plane-wave formulation as in Eq. (5), which leads to the ear-alignment formulation:

$$h_a^{L\backslash R}(k,\Omega) = h^{L\backslash R}(k,\Omega)e^{-ikr_a \cos\Theta_{L\backslash R}}, \tag{11}$$

where $\Theta_{L\backslash R}$ is the angle between the direction of the ear, $\Omega_{L\backslash R}$, and the direction of the sound source, Ω, and $\cos\Theta_{L\backslash R} = \cos\theta \cos\theta_{L\backslash R} + \cos\left(\phi - \phi_{L\backslash R}\right)\sin\theta \sin\theta_{L\backslash R}$. It is important to note that this ear-alignment process is invertible, which means that going from $h^{L\backslash R}$ to $h_a^{L\backslash R}$ and back can be performed without any loss of information.

Figure 2 presents an example of the SH spectrum of a KEMAR HRTF [26, 52], for the basic and ear-aligned HRTF representations. The SH spectrum, which is the energy of the SH coefficients at every order n, is computed as:

$$E_n(f) = \sum_{m=-n}^{n} |h_{nm}(f)|^2, \tag{12}$$

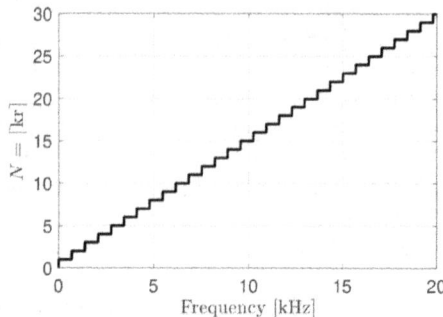

Figure 1.
Added SH order due to the dual-centering of the HRTF, $N = \lceil kr_a \rceil$, as a function of frequency ($\lceil \cdot \rceil$ is the ceiling function). Computed for the free-field HRTF with $r_a = 8$ cm and $c = 343$ m/s.

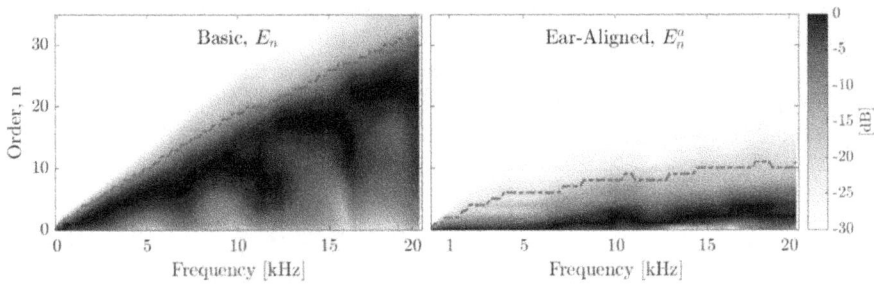

Figure 2.
Normalized SH spectra, E_n, of basic and ear-aligned KEMAR HRTF representations, computed according to Eq. (12). The dashed gray line represents the order at which 99% of the energy is contained.

and normalized by the maximum value for each frequency. The figure shows how the energy of the high-order SH coefficients of the ear-aligned HRTF is significantly reduced compared to the basic HRTF. This validates the finding from Section 3.1, in which the high orders of the basic HRTF actually originate from the translation from the origin. In particular, the order at which 99% of the energy is contained is reduced to be below order 10 for all frequencies.

It is interesting to note that the SH order reduction of the ear-aligned HRTF can explain the reduced order of the time-alignment method. This is discussed in detail in [26, 27]. The ear-alignment can be interpreted as" virtually" removing the inherent delay in an HRTF caused by normalizing the pressure at the ear by the pressure at the origin. This is evident from Eq. (11), where the phase in the exponent represents a delay from the origin to the ear due to a source at Ω. The main difference between the time-alignment and ear-alignment methods is as follows. Performing time-alignment requires numerical estimation of the time delays; this may be challenging and its accuracy may depend on the HRTF direction and on the quality of the measurements [53, 54]. In contrast, ear-alignment can be performed parametrically with the parameters r_a and $\Omega_{L\backslash R}$. Moreover, using the ear-alignment with fixed parameters makes it data-independent, which improves its robustness to measurement noise (as discussed comprehensively in [29]).

4. Binaural reproduction based on bilateral ambisonics and ear-aligned HRTFs

While the ear-alignment method leads to efficient SH representation of the HRTF, incorporating the pre-processed ear-aligned HRTF in a binaural reproduction process is not trivial. The computation of the binaural signal (Eq. (3)) requires the HRTF and the Ambisonics signals to be represented in the same coordinate system and around the same origin. One way to align them is to re-synthesize the HRTF phase before the computation of the binaural signal, which will increase its order back to the original high order, and will cause similar truncation error to that in the Basic Ambisonics reproduction. Another way is to use the MagLS approach, which completely ignores the HRTF phase component at high frequencies [31]. Alternatively, the Binaural B-Format approach, presented by Jot *et al.* [38], can be used. In this approach, two B-Format recordings at the ear locations are used, together with a minimum-phase approximation of the HRTF and an ITD estimation based on a spherical head model. The Binaural B-format can be extended by using the ear-aligned HRTF together with high-order Ambisonics signals that are defined around the ear locations. This approach is denoted as Bilateral Ambisonics reproduction [41, 42].

Assuming that the plane-wave amplitude density function, denoted by $a^{L\backslash R}(k, \Omega)$, is given at the position of the ear, then the binaural signal can be computed directly at the listener's ears, using the ear-aligned HRTF, similarly to in Eq. (1):

$$p^{L\backslash R}(k) = \int_{\Omega} a^{L\backslash R}(k, \Omega) h_a^{L\backslash R}(k, \Omega) \Omega. \tag{13}$$

From here, the Bilateral Ambisonics reproduction of order N can be formulated as:

$$p^{L\backslash R}(k) = \sum_{n=0}^{N} \sum_{m=-n}^{n} \left[\tilde{a}_{nm}^{L\backslash R}(k) \right]^* h_{a\,nm}^{L\backslash R}(k), \tag{14}$$

where $\tilde{a}_{nm}^{L\backslash R}(k)$ and $h_{a\,nm}^{L\backslash R}(k)$ are the SH coefficients of $\left[a^{L\backslash R}(k, \Omega) \right]^*$ and $h_a^{L\backslash R}(k, \Omega)$, respectively. It is important to note that, in contrast to $a(k, \Omega)$, which is the plane-wave amplitude density function of the sound-field as observed at the position of the center of the head, $a^{L\backslash R}(k, \Omega)$ is observed at the position of the ears. **Figure 3** demonstrates the differences between the two coordinate systems. The standard coordinate system, denoted by black dashed axes with its origin at the center of the head, is used for the computations of the binaural signals in Eqs. (1) and (3) using the Basic Ambisonics signals, $\tilde{a}_{nm}(k)$, for both ears. The bilateral coordinate systems, denoted by red dotted axes with their origin at the positions of the ears, are used for the computation in Eq. (14) using the Bilateral Ambisonics signals, $\tilde{a}_{nm}^{L\backslash R}(k)$, which are different for each ear. **Figure 4** demonstrates the signal-flow of the Basic and Bilateral Ambisonics.

Theoretically, the plane-wave amplitude density function at the position of the ear can be computed from the center function by translation of the sound-field [46], which is computed as $a^{L\backslash R}(k, \Omega) = a(k, \Omega) e^{ikr_a \cos \Theta_{L\backslash R}}$; however, this will lead to equivalence between Eq. (13) and Eq. (1), which means that the binaural signals

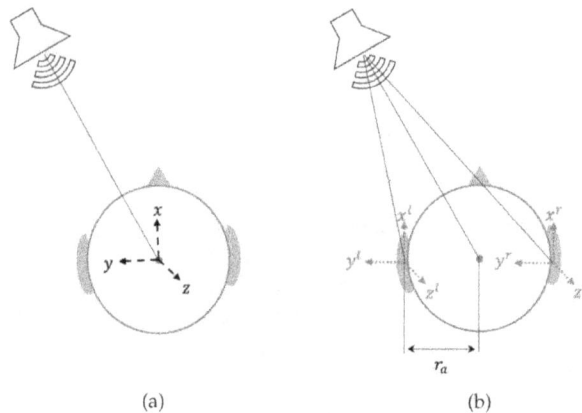

(a)   (b)

Figure 3.
Diagram of the standard (a) and Bilateral (b) coordinate systems. The origin of the standard coordinate system is at the center of the head, while in the bilateral coordinate system the origin is at the position of the ear.

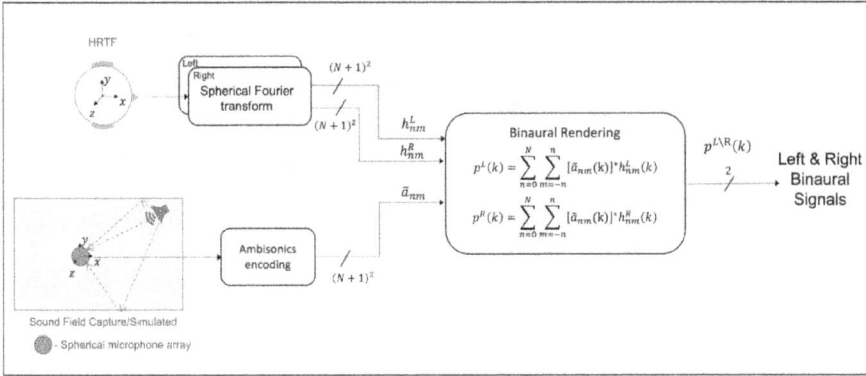

(a)

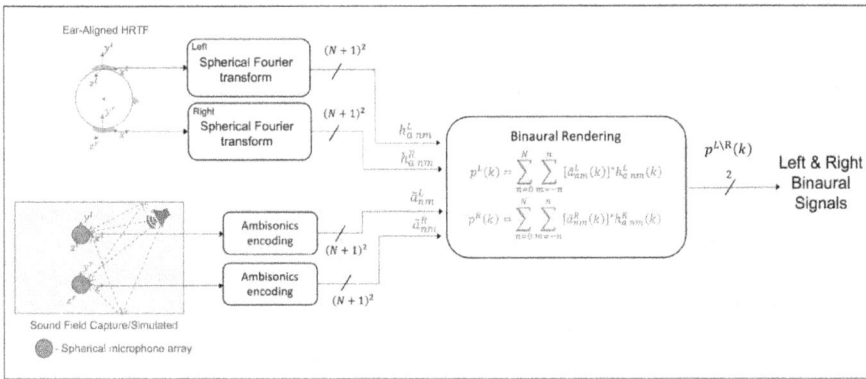

(b)

Figure 4.
Binaural reproduction signal-flow of the Basic (a) and Bilateral (b) Ambisonics.

will be identical. Thus, the same truncation error as in the Basic Ambisonics reproduction is introduced. Alternatively, if a low-order plane-wave amplitude density function is given directly at the position of the ear, the Bilateral Ambisonics-based signals (from Eq. (14)) may potentially be more accurate than the Basic Ambisonics reproduction (from Eq. (3)) due to the lower-order nature of the ear-aligned HRTF compared to the unprocessed basic HRTF.

Figure 5 demonstrates the improved accuracy of the Bilateral Ambisonics reproduction. The figure shows the magnitude response of the binaural signals for a single plane-wave of unit amplitude arriving from direction $(\theta, \phi) = (90°, 20°)$, using a KEMAR HRTF, with $N = 1, 4$, and a high-order reference of $N = 40$. For the low-order signals computed using Basic Ambisonics reproduction, a high-frequency roll-off above the sphere cut-off frequency, $kr_a = N$ [14], is clearly observed. This is discussed further in [19]. Additionally, amplitude distortion is also observed at these high frequencies. The Bilateral Ambisonics-based signals seem significantly more accurate in terms of both frequency roll-off and distortion, where reproduction of order $N = 4$ seems to preserve the signal magnitude up to almost 20 kHz, including the important spectral cues (peaks and notches). Further evaluation of the performance of the Bilateral Ambisonics reproduction is presented in Section 6.

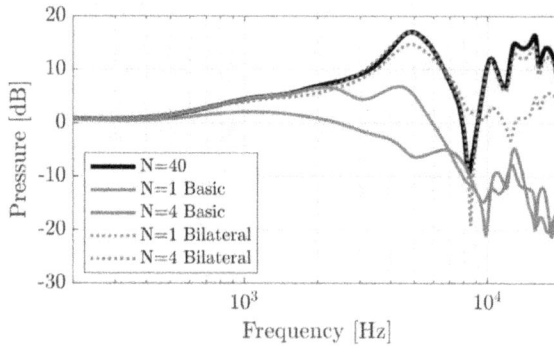

Figure 5.
Magnitude of a left ear binaural signal of a single plane-wave from direction $(\theta, \phi)=(90°, 20°)$, with HRTF of KEMAR, computed with Basic Ambisonics reproduction (solid lines) and with Bilateral Ambisonics reproduction (dashed lines), with $N = 1, 4$, compared to a high-order reference with $N = 40$.

5. Head-tracking in bilateral ambisonics reproduction

While Bilateral Ambisonics leads to a more efficient representation of the spatial audio signal and more accurate binaural reproduction, such a procedure will result in a static binaural reproduction. In contrast to the Basic Ambisonics reproduction, where head-rotations can be incorporated in post-processing by a simple rotation of the Ambisonics signals using Wigner-D functions [55], performing this operation in Bilateral Ambisonics is not straightforward. A method to incorporate head-rotations in Bilateral Ambisonics reproduction is presented in this section.

Consider the specific case where a binaural signal is played via headphones to a listener, representing a spatial acoustic scene composed of a single sound source. According to the Bilateral Ambisonics format, the scene is represented by two Ambisonics signals with their origin at the listener's expected ear positions, as seen in **Figure 6a**. Note that the microphone symbols in **Figure 6** represent the left and right Ambisonics signals origin. Upon playback of the acoustic scene, the listener is expected to perceive a virtual source from the direction of the real source (in this example about 30° to the left). Next, the listener rotates his/her head while

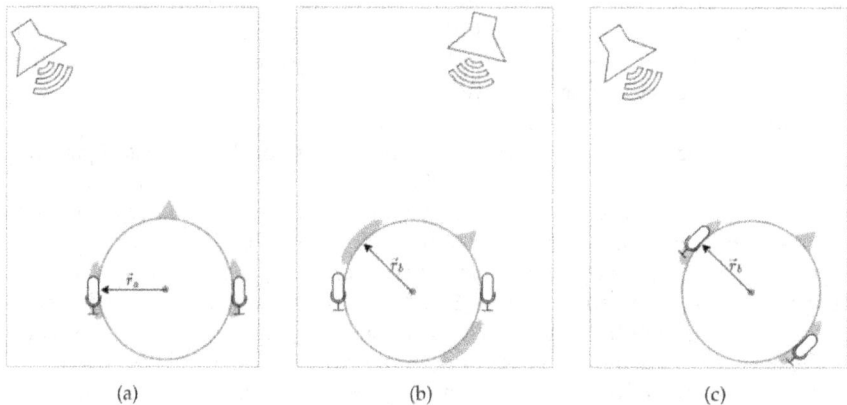

(a)   (b)   (c)

Figure 6.
Demonstration of the head-tracking method in bilateral coordinate systems, (a) before the head-rotation, (b) after head-rotation and without head-tracking and (c) after head-rotation and with head-tracking. $\vec{r}_a$ and $\vec{r}_b$ are the head-rotation ear vectors before and after head rotation, respectively.

listening; this action will result in the virtual source changing its position in space, remaining at about 30° to the left, as illustrated in **Figure 6b**. One way to compensate for the head rotation is to acquire new Bilateral Ambisonics signals located at the listener ears' new locations, and also to rotate them according to the angle of rotation of the listener's head, as illustrated in **Figure 6c**. This, of course, may not be a practical option since acquiring new Bilateral Ambisonics signals requires re-synthesizing the sound-field, in the case of a simulation, or adjusting the position of the physical microphone arrays, in the case of sound field capture. The former may be computationally expensive, while the latter is practically infeasible since recording is typically performed independently from the listener's head orientation. Note that a multi-microphone binaural recording method could be employed, similar to the Motion-Tracked Binaural recording method [56], though this is solution will be complex in terms of the recording resources. Hence, developing methods that compensate for the listener head movements using head-tracking is of great importance for Bilateral Ambisonics recording and reproduction. **Figure 6** shows that as a result of the rotation in this case, the ears (which are the Ambisonics reference point) change their orientation while also translating in space. Proper compensation for head-rotation needs to take both movements into account.

Now, consider the general case, where an arbitrary sound-field is represented by a plane-wave amplitude density function, denoted by $a(k, \Omega)$, given at the position of the ear. Note that $a(k, \Omega)$ represents the same function as $a^{L\backslash R}(k, \Omega)$ from Eq. (13), but the superscript $^{L\backslash R}$ is left out for notation simplicity since the operation is similar for both ears. Assuming that the ear position with respect to the head center is known before and after the rotation, denoted as $\vec{r}_a$ and $\vec{r}_b$, respectively, head-tracking can be performed by translation of the plane-wave amplitude density function $a(k, \Omega)$, accordingly. This translation can be performed by a phase-shifting operation, as follows [46, 57]:

$$a^t(k, \Omega) = a(k, \Omega)e^{-i\vec{k}\cdot(\vec{r}_b - \vec{r}_a)} = a(k, \Omega)e^{-ikr_a(\cos\Theta_a - \cos\Theta_b)}, \quad (15)$$

where $a^t(k, \Omega)$ is the translated plane-wave amplitude density function, which represents the plane-wave amplitude density around the ear of the listener after head-rotation (but with the pre-rotation orientation). $r_a = |\vec{r}_a|$ is the head radius, $\vec{k}$ is the wave vector, Θ_a is the angle between the sound source direction, Ω, and the pre-rotation ear position, $\vec{r}_a$, and Θ_b is the angle between Ω and $\vec{r}_b$. **Figure 7**

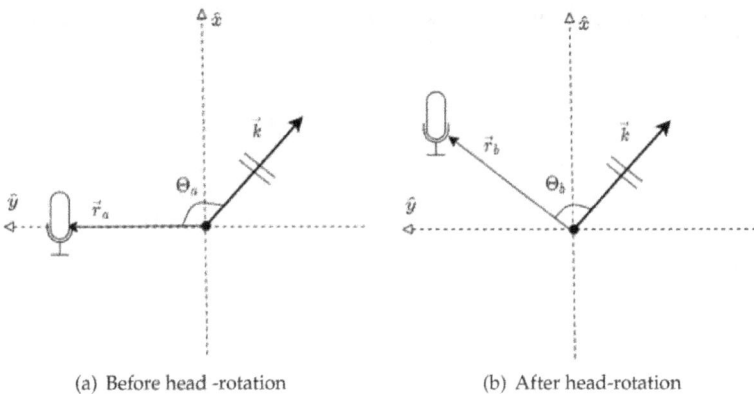

(a) Before head -rotation   (b) After head-rotation

Figure 7.
Schematic illustration for left ear microphone array translation due to head rotation: $\vec{r}_a$ is the left ear vector with respect to the head center, $\vec{r}_b$ is the left ear vector after a clock-wise rotation, $\vec{k}$ is the wave vector of the plane-wave, where Θ_a and Θ_b represent the angle between the ear vector and the wave vector.

demonstrates this translation for a simple case where the sound-field is comprised of a single plane-wave and the microphone symbols represent the measurement position of $a(k, \Omega)$ made by microphone arrays.

Next, the orientation of the translated plane-wave amplitude density function is corrected by applying rotation. This is formulated in the SH domain by:

$$a_{nm}^r (k) = \sum_{m'=-n}^{n} a_{nm'}^t (k) D_{mm'}^n (\alpha, \beta, \gamma), \tag{16}$$

where $a_{nm}^t(k)$ are the SH coefficients of $a^t(k, \Omega)$, $D_{mm'}^n$ denotes the Wigner D functions, and a_{nm}^r are the rotated Bilateral Ambisonics signals. (α, β, γ) are the Euler angles [58] of the head-rotation, which are assumed to be known, for example from a head-tracker. Note that this procedure needs to be applied to both left and right ears.

In practice, the Bilateral Ambisonics signals will be order limited due to the constraints mentioned in Section 2. The finite order representation, in turn, leads to limitations in the accuracy of the suggested method. These limitations will be presented and demonstrated in numerical simulations in Section 6.

6. Performance analysis

This section presents an objective evaluation of the performance of the proposed Bilateral Ambisonics reproduction approach, and compares it to that of the Basic Ambisonics+MagLS reproduction method.

A binaural signal for a sound-field composed of a single plane-wave of unit amplitude, as presented in **Figure 5**, is computed, and the Normalized Mean Square Error (NMSE) for the left ear is evaluated as:

$$\varepsilon^L(f) = 10 \log_{10} \frac{\left| p_{ref}^L(f) - p^L(f) \right|^2}{\left| p_{ref}^L(f) \right|^2}, \tag{17}$$

where p_{ref}^L is the reference high-order binaural signal computed using Eq. (3) with $N = 40$, and p^L is the binaural signal computed using Eq. (3) or (14). The NMSE, although positive and real, is sensitive to both the magnitude and the phase

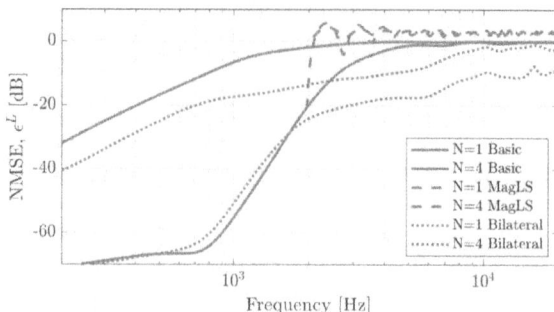

Figure 8.
NMSE of binaural signals computed for sound-fields composed of a single plane-wave, averaged over 434 plane-wave directions (distributed according to a Lebedev grid), with HRTF of KEMAR. The NMSE is computed using Eq. (17), with Basic Ambisonics reproduction (solid lines), with Basic Ambisonics reproduction with MagLS [31] (dashed lines), and with Bilateral Ambisonics reproduction (dotted lines), with $N = 1, 4$, and a high-order reference with $N = 40$.

errors in the binaural signal. The NMSE is averaged over a range of 434 plane-waves with incidence angles distributed nearly-uniformly over the sphere, using the Lebedev sampling scheme of order 17 [59].

Figure 8 shows this averaged NMSE. For the MagLS approach, a cutoff frequency of 2 kHz was used, as indicated by the increased error above this frequency, where the phase is completely inaccurate. The figure demonstrates the improvement in the accuracy of the Bilateral Ambisonics reproduction, compared to the Basic Ambisonics reproduction methods, where at high frequencies, up to about 5 kHz for $N = 1$ and 15 kHz for $N = 4$, the errors are lower by 10–20 dB. Two important spatial cues for sound source localization are the Interaural Time Difference (ITD) and Interaural Level Difference (ILD). Both were shown to be affected by the truncation error due to low-order reproduction [29]. **Figure 9** shows the ITDs, ILDs and their corresponding errors relative to a high-order reference ($N = 40$). The ITDs and ILDs were computed for binaural signals of a single plane-wave sound-field with incident angles across the left horizontal half-plane ($\theta = 90°$; $0° \leq \phi \leq 180°$) with 1° resolution, and with a KEMAR HRTF. The ITDs were esti-

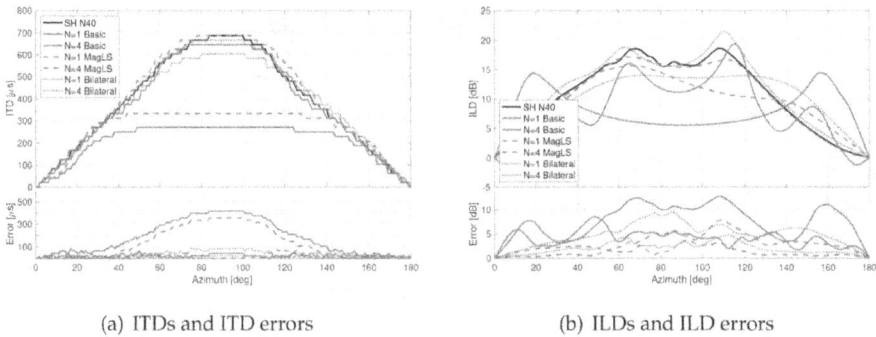

(a) ITDs and ITD errors   (b) ILDs and ILD errors

Figure 9.
ITDs, ILDs and their corresponding errors as a function of azimuth angle for binaural signals computed for sound-fields composed of a single plane-wave from 180 directions on the left horizontal plane (the right side is symmetrical), with HRTF of KEMAR. The signals were computed using Basic Ambisonics reproduction with and without MagLS, and Bilateral Ambisonics reproduction.

mated using the onset detection method, applied to a 2 kHz low-pass filtered version of the signals [54]. The ILDs were calculated and averaged across 18 auditory filter bands as [5]:

$$\mathrm{ILD}(f_c, \Omega) = 10 \log_{10} \frac{\int C(f, f_c) |p^L(f)|^2 df}{\int C(f, f_c) |p^R(f)|^2 df}, \tag{18}$$

$$\mathrm{ILD}_{av}(\Omega) = \frac{1}{18} \sum_{f_c} \mathrm{ILD}(f_c, \Omega), \tag{19}$$

where C is a Gammatone filter with center frequency f_c, as implemented in the Auditory Toolbox [60]. The integral is evaluated between 1.5kHz and 16kHz and f_c is restricted accordingly. This computation facilitates a perceptually motivated smoothing of the ILD across frequencies, which is required for appropriate comparison between ILDs.

Comparison of the ITD errors with the Just Notable Differences (JND) values reported by Andreopoulou and Katz in [54] (40 μs for the frontal directions and about 100 μs for the lateral directions) reveals the main advantage of the Bilateral

Ambisonics approach, where the phase information is preserved and the ITD errors are below the JND even at $N = 1$.

Figure 9b shows that both the MagLS and the Bilateral approaches achieve significant improvement in the ILD accuracy compared to the Basic Ambisonics reproduction. While with the Basic Ambisonics reproduction the ILD errors are above the JND ($\sim$1 dB [61, 62]) even with $N = 4$, with the MagLS and the Bilateral Ambisonics reproduction the errors for $N = 4$ are below the JND for most angles. Relatively high errors can be seen at the lateral angles compared to the front and back directions. This can be explained by the fact that the ILD at the front and back directions is close to zero, where the errors are expected to be small due to the symmetry of the HRTF model. Nevertheless, both the MagLS and the Bilateral Ambisonics reproduction led to substantially lower ILD errors compared to the Basic Ambisonics reproduction.

As discussed in Section 5, a limitation of the Bilateral Ambisonics method compared to Basic Ambisonics is found in terms of the incorporation of head-tracking in post-processing. In Section 5, a method to overcome this limitation was suggested. To evaluate the performance of this method, a simulation study was conducted. The simulation results aim to evaluate the NMSE introduced by the head rotation and its dependence on the Bilateral Ambisonics signal order and the head rotation angle. In the simulation, a head was positioned in free-field, facing the $\hat{x}$ direction with the ears positioned on the xy plane. A sound-field was generated, consisting of a single plane-wave with unit amplitude arriving from directions that are taken from the Lebedev sampling scheme, using the same sampling scheme mentioned earlier. The Bilateral Ambisonics signal, $a_{nm}^L(k)$, is calculated with respect to the left ear position $\vec{r}_a$ up to order N. Note that the superscript ^{L} denoting the left ear will be removed for brevity from now on. The signal is then transformed to $a(k, \Omega)$ with the discrete inverse spherical Fourier transform (DISFT) [46]. Next, the head is rotated by γ degrees clockwise in the horizontal plane, as shown in **Figure 6b**, resulting in a new rotated left ear position $\vec{r}_b$. The translated plane-wave amplitude density function, $a^t(k, \Omega)$, is computed using Eq. (15). Next, Eq. (16) is used to calculate $a_{nm}^r(k)$ from $a_{nm}^t(k)$, the discrete spherical Fourier transform (DSFT) [46] of $a^t(k, \Omega)$. The signal $a_{nm}^r(k)$ represents the head-rotated left ear Bilateral Ambisonics signal; note that the right ear signal can be calculated in a similar manner. Finally, the left ear binaural signal with head-tracked Bilateral Ambisonics, $p(f)$, is calculated using Eq. (14) with $a_{nm}^r(k)$ and a KEMAR HRTF. The reference binaural signal, $p_{ref}(f)$, is calculated using Eq. (14) with an accurately generated Bilateral Ambisonics signal $a_{nm}^{ref}(k)$ of order N at the head-rotated position. The NMSE is calculated using Eq. (17), and averaged over the 434 sampling scheme directions.

Figure 10a shows the NMSE between $p_{ref}(f)$ and $p(f)$ for a head rotation of $\gamma = 30°$ and different reproduction orders, $N = 1, 4, 10$. The figure demonstrates the improvement in the NMSE as the order increases. Additionally, the figure demonstrates how the error increases with frequency. For $N = 1, 4, 10$ an error of less then -10 dB is achieved up to about 1, 5 kHz and 11 kHz, respectively. This result indicates that, for example, with order $N = 4$ and a rotation angle of 30° the suggested rotation method will experience a noticeable loss in accuracy above 5kHz, compared to the reference. To evaluate the performance of the suggested method for different head rotation angles, the order was kept at $N = 4$ and various values of head rotation angle, γ, were used. **Figure 10b** illustrates how the performance deteriorates as the rotation angle increases. For $\gamma = 30°, 60°, 90°$ an error of less than -10 dB is achieved up to about 5kHz, 3kHz and 2kHz, respectively.

We now compare between binaural reproduction performance with head-tracked Bilateral Ambisonics, head-tracked MagLS and with head-tracked Basic Ambisonics.

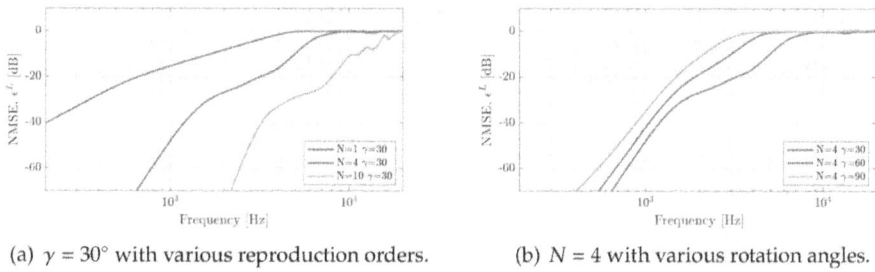

(a) $\gamma = 30°$ with various reproduction orders.　　(b) $N = 4$ with various rotation angles.

Figure 10.
NMSE of binaural signals computed using Bilateral Ambisonics reproduction with head rotation as in Eq. (16), for various orders (a) and rotation angles (b).

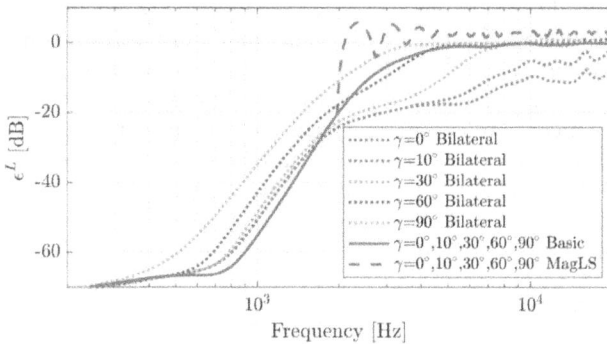

Figure 11.
NMSEs of binaural signals computed using rotated Basic, MagLS and Bilateral Ambisonics signals with order $N = 4$ relative to a high-order Basic Ambisonics reproduction with $N = 40$, with various rotation angles, γ. The NMSE is averaged over 434 plane-wave directions.

In the simulation (which is identical to the previously described simulation), the NMSE is measured for head-tracked binaural signals computed using Basic, MagLS and Bilateral Ambisonics reproductions with order $N = 4$, and compared with a high-order reference computed using Basic Ambisonics reproduction with order $N = 40$. The head-tracked Bilateral Ambisonics signals are calculated with the suggested method, using Eqs. (15) and (16), and both the head-tracked Basic Ambisonics and MagLS signals are calculated in the SH domain using Eq. (16). Note that for head-tracking with Basic Ambisonics and MagLS, the error is independent of the rotation angle, γ. **Figure 11** presents the results for different head-rotation angles, γ. As expected, the rotation procedure compromises the accuracy of the binaural signal with Bilateral Ambisonics at high frequencies. For $\gamma = 10°$, the Bilateral Ambisonics reproduction retains its advantage in accuracy compared to the Basic Ambisonics reproduction up to around 20 kHz. However, for a head-rotation of $\gamma = 30°$, the Bilateral Ambisonics reproduction retains its advantage only up to about 7 kHz. For a head-rotation of $\gamma = 60°$, the two reproduction schemes are equally accurate. For a head-rotation of $\gamma = 90°$, the Bilateral Ambisonics reproduction results in an error of less than -10 dB up to about 2 kHz, compared to 2.5 kHz for Basic Ambisonics. Similar behavior was also observed for other reproduction orders. These results indicate that, in this case, the suggested rotation method is mainly beneficial for head rotations up to 60°. Note that 60° means that the listener can turn his/her head 60° both to the left and to the right. The inaccuracies depicted in **Figures 10** and **11** relating to the reproduction order N and head-rotation angle γ, can be explained by errors due to the translation operation described in Eq. (15) [46, 57].

Further evaluation of head-tracking compensation is the subject of ongoing research. The study could include evaluation of ITD/ILD reconstruction, Lateral Error, Polar error in median plane, Coloration error [26] and subjective listening tests.

7. Conclusions

This chapter presented a detailed description of the Bilateral Ambisonics reproduction method. The method incorporates a pre-processed ear-aligned HRTF, which provides an efficient representation of the HRTF in the SH domain, with bilateral representation of the Ambisonics signals. The method was shown to improve the accuracy of low-order binaural reproduction in comparison to Basic Ambisonics reproduction in terms of reduced errors in the binaural signals, as well as more accurate ITD and ILD. The two main limitations of this method are the requirement for two Ambisonics signals at the positions of the ears, and the difficulty of incorporating head-tracking. The latter has been addressed in this chapter by presenting a method to incorporate head-tracking in post-processing. Ways should be sought to mitigate the requirement for two different Ambisonics signals, for example by transforming a Basic Ambisonics signal into a Bilateral Ambisonics signal.

Conflict of interest

The authors declare no conflict of interest.

Author details

Zamir Ben-Hur[1]*, David Alon[1], Or Berebi[2], Ravish Mehra[1] and Boaz Rafaely[2]

1 Facebook Reality Labs Research, Facebook, USA

2 School of Electrical and Computer Engineering, Ben-Gurion University of the Negev, Israel

*Address all correspondence to: zamirbh@fb.com

IntechOpen

References

[1] Begault DR. 3–D sound for virtual reality and multimedia. NASA, Ames Research Center, Moffett Field, California. 2000:132–136.

[2] Vorländer M. Auralization: fundamentals of acoustics, modeling, simulation, algorithms and acoustic virtual reality. Springer Science & Business Media; 2007.

[3] Blauert J. Spatial hearing: the psychophysics of human sound localization. MIT press; 1997.

[4] Møller H. Fundamentals of binaural technology. Applied acoustics. 1992;36 (3):171–218.

[5] Xie B. Head-related transfer function and virtual auditory display. J. Ross Publishing; 2013.

[6] Brandstein M, Ward D. Microphone arrays: signal processing techniques and applications. Springer Science & Business Media; 2013.

[7] Duraiswami R, Zotkin DN, Li Z, Grassi E, Gumerov NA, Davis LS. High order spatial audio capture and binaural head-tracked playback over headphones with HRTF cues. 119th Convention of Audio Engineering Society. 2005.

[8] Sheaffer J, Van Walstijn M, Rafaely B, Kowalczyk K. Binaural reproduction of finite difference simulations using spherical array processing. IEEE/ACM Transactions on Audio, Speech, and Language Processing. 2015;23(12):2125–2135.

[9] Gerzon MA. Ambisonics in multichannel broadcasting and video. Journal of the Audio Engineering Society. 1985;33(11):859–871.

[10] Rafaely B, Avni A. Interaural cross correlation in a sound field represented by spherical harmonics. The Journal of the Acoustical Society of America. 2010; 127(2):823–828.

[11] Zotter F, Frank M. Ambisonics: A practical 3D audio theory for recording, studio production, sound reinforcement, and virtual reality. Springer Nature; 2019.

[12] Jeffet M, Shabtai NR, Rafaely B. Theory and perceptual evaluation of the binaural reproduction and beamforming tradeoff in the generalized spherical array beamformer. IEEE/ACM Transactions on Audio, Speech, and Language Processing. 2016;24(4):708–718.

[13] Alon DL, Rafaely B. Beamforming with optimal aliasing cancelation in spherical microphone arrays. IEEE/ ACM Transactions on Audio, Speech, and Language Processing. 2015;24(1): 196–210.

[14] Rafaely B. Plane-wave decomposition of the sound field on a sphere by spherical convolution. The Journal of the Acoustical Society of America. 2004;116(4):2149–2157.

[15] Rafaely B. Analysis and design of spherical microphone arrays. Speech and Audio Processing, IEEE Transactions on. 2005;13(1):135–143.

[16] Noisternig M, Sontacchi A, Musil T, Holdrich R. A 3D ambisonic based binaural sound reproduction system. Journal of the Audio Engineering Society. 2003 June.

[17] Zhang W, Abhayapala TD, Kennedy RA, Duraiswami R. Insights into head-related transfer function: Spatial dimensionality and continuous representation. The Journal of the Acoustical Society of America. 2010;127 (4):2347–2357.

[18] Avni A, Ahrens J, Geier M, Spors S, Wierstorf H, Rafaely B. Spatial

perception of sound fields recorded by spherical microphone arrays with varying spatial resolution. The Journal of the Acoustical Society of America. 2013;133(5):2711–2721.

[19] Ben-Hur Z, Brinkmann F, Sheaffer J, Weinzierl S, Rafaely B. Spectral equalization in binaural signals represented by order-truncated spherical harmonics. The Journal of the Acoustical Society of America. 2017;141 (6):4087–4096.

[20] Ben-Hur Z, Alon DL, Rafaely B, Mehra R. Loudness stability of binaural sound with spherical harmonic representation of sparse head-related transfer functions. EURASIP Journal on Audio, Speech, and Music Processing. 2019 Mar;2019(1):5.

[21] Ahrens J, Andersson C. Perceptual evaluation of headphone auralization of rooms captured with spherical microphone arrays with respect to spaciousness and timbre. The Journal of the Acoustical Society of America. 2019; 145(4):2783–2794.

[22] Politis A, McCormack L, Pulkki V. Enhancement of ambisonic binaural reproduction using directional audio coding with optimal adaptive mixing. In: 2017 IEEE Workshop on Applications of Signal Processing to Audio and Acoustics (WASPAA). IEEE; 2017. p. 379–383.

[23] Politis A, Tervo S, Pulkki V. COMPASS: Coding and Multidirectional Parameterization of Ambisonic Sound Scenes. In: 2018 IEEE International Conference on Acoustics, Speech and Signal Processing (ICASSP); 2018. p. 6802–6806.

[24] Mccormack L, Politis A. SPARTA & COMPASS: Real-time implementations of linear and parametric spatial audio reproduction and processing methods. Journal of the Audio Engineering Society. 2019 March.

[25] Barrett N, Berge S. A new method for B-format to binaural transcoding. Journal of the Audio Engineering Society. 2010 October.

[26] Brinkmann F, Weinzierl S. Comparison of head-related transfer functions pre-processing techniques for spherical harmonics decomposition. Journal of the Audio Engineering Society. 2018 August.

[27] Evans MJ, Angus JA, Tew AI. Analyzing head-related transfer function measurements using surface spherical harmonics. The Journal of the Acoustical Society of America. 1998;104 (4):2400–2411.

[28] Romigh GD, Brungart DS, Stern RM, Simpson BD. Efficient real spherical harmonic representation of head-related transfer functions. IEEE Journal of Selected Topics in Signal Processing. 2015;9(5):921–930.

[29] Ben-Hur Z, Alon DL, Mehra R, Rafaely B. Efficient Representation and Sparse Sampling of Head-Related Transfer Functions Using Phase-Correction Based on Ear Alignment. IEEE/ACM Transactions on Audio, Speech, and Language Processing. 2019; 27(12):2249–2262.

[30] Zaunschirm M, Schörkhuber C, Höldrich R. Binaural rendering of Ambisonic signals by head-related impulse response time alignment and a diffuseness constraint. The Journal of the Acoustical Society of America. 2018; 143(6):3616–3627.

[31] Schörkhuber C, Zaunschirm M, Höldrich R. Binaural rendering of ambisonic signals via magnitude least squares. In: Proceedings of the DAGA. vol. 44; 2018. p. 339–342.

[32] Wightman FL, Kistler DJ. The dominant role of low-frequency interaural time differences in sound localization. The Journal of the

Acoustical Society of America. 1992;91 (3):1648–1661.

[33] Macpherson EA, Middlebrooks JC. Listener weighting of cues for lateral angle: the duplex theory of sound localization revisited. The Journal of the Acoustical Society of America. 2002;111 (5):2219–2236.

[34] Minnaar P, Christensen F, Moller H, Olesen SK, Plogsties J. Audibility of all-pass components in binaural synthesis. Journal of the Audio Engineering Society. 1999 may.

[35] Benichoux V, Rébillat M, Brette R. On the variation of interaural time differences with frequency. The Journal of the Acoustical Society of America. 2016;139(4):1810–1821.

[36] Lübeck T, Helmholz H, Arend JM, Pörschmann C, Ahrens J. Perceptual Evaluation of Mitigation Approaches of Impairments due to Spatial Undersampling in Binaural Rendering of Spherical Microphone Array Data. Journal of the Audio Engineering Society. 2020;68(6):428–440.

[37] Hold C, Gamper H, Pulkki V, Raghuvanshi N, Tashev IJ. Improving binaural ambisonics decoding by spherical harmonics domain tapering and coloration compensation. In: ICASSP 2019–2019 IEEE International Conference on Acoustics, Speech and Signal Processing (ICASSP). IEEE; 2019. p. 261–265.

[38] Jôt JM, Wardle S, Larcher V. Approaches to binaural synthesis. Journal of the Audio Engineering Society. 1998 september.

[39] Jot JM, Larcher V, Pernaux JM. A comparative study of 3-d audio encoding and rendering techniques. Journal of the Audio Engineering Society. 1999 march.

[40] Larcher V, Warusfel O, Jot JM, Guyard J. Study and comparison of

efficient methods for 3-d audio spatialization based on linear decomposition of hrtf data. Journal of the Audio Engineering Society. 2000 february.

[41] Ben-Hur Z, Alon D, Mehra R, Rafaely B. Binaural reproduction using bilateral ambisonics. Journal of the Audio Engineering Society. 2020 august.

[42] Ben-Hur Z, Alon DL, Mehra R, Rafaely B. Binaural Reproduction Based on Bilateral Ambisonics and Ear-Aligned HRTFs. IEEE/ACM Transactions on Audio, Speech, and Language Processing. 2021;29:901–913.

[43] Park M, Rafaely B. Sound-field analysis by plane-wave decomposition using spherical microphone array. The Journal of the Acoustical Society of America. 2005;118(5):3094–3103.

[44] Ben-Hur Z, Sheaffer J, Rafaely B. Joint sampling theory and subjective investigation of plane-wave and spherical harmonics formulations for binaural reproduction. Applied Acoustics. 2018;134:138–144.

[45] mh acoustics. em32 Eigenmike microphone array release notes; 2009. 25 Summit Ave Summit NJ 07901, h ttp://www.mhacoustics.com/produc ts#eigenmike1.

[46] Rafaely B. Fundamentals of Spherical Array Processing. vol. 8. Springer; 2015.

[47] Pörschmann C, Arend JM, Brinkmann F. Directional Equalization of Sparse Head-Related Transfer Function Sets for Spatial Upsampling. IEEE/ACM Transactions on Audio, Speech, and Language Processing. 2019; 27(6):1060–1071.

[48] Ben-Hur Z, Alon DL, Mehra R, Rafaely B. Sparse Representation of HRTFs by Ear Alignment. In: 2019 IEEE Workshop on Applications of Signal

Processing to Audio and Acoustics (WASPAA). IEEE; 2019. p. 70–74.

[49] Williams EG. Fourier acoustics: sound radiation and nearfield acoustical holography. Academic press; 1999.

[50] Arfken GB, Weber HJ, Harris FE. Mathematical Methods for Physicists: A Comprehensive Guide. Elsevier; 2012. Available from: https://books. google. com/books?id = qLFo Z-PoGIC.

[51] Ward DB, Abhayapala TD. Reproduction of a plane-wave sound field using an array of loudspeakers. IEEE Transactions on speech and audio processing. 2001;9(6):697–707.

[52] Dinakaran M, Brinkmann F, Harder S, Pelzer R, Grosche P, Paulsen RR, et al. Perceptually motivated analysis of numerically simulated head-related transfer functions generated by various 3D surface scanning systems. In: 2018 IEEE International Conference on Acoustics, Speech and Signal Processing (ICASSP). IEEE; 2018. p. 551–555.

[53] Katz BF, Noisternig M. A comparative study of interaural time delay estimation methods. The Journal of the Acoustical Society of America. 2014;135(6):3530–3540.

[54] Andreopoulou A, Katz BF. Identification of perceptually relevant methods of inter-aural time difference estimation. The Journal of the Acoustical Society of America. 2017;142 (2):588–598.

[55] Rafaely B, Kleider M. Spherical microphone array beam steering using Wigner-D weighting. IEEE Signal Processing Letters. 2008;15:417–420.

[56] algazi vr, duda ro, thompson dm. motion-tracked binaural sound. Journal of the audio engineering society. 2004 november;52(11):1142–1156.

[57] Berebi O, Ben-Hur Z, Alon D, Rafaely B. Enabling Head-Tracking for Binaural Sound Reproduction Based on Bilateral Ambisonics. International Conference on Immersive and 3D Audio (I3DA). 2021.

[58] Arfken GB, Weber HJ. Mathematical methods for physicists. American Association of Physics Teachers; 1999.

[59] Lecomte P, Gauthier PA, Langrenne C, Garcia A, Berry A. On the use of a Lebedev grid for Ambisonics. Journal of the Audio Engineering Society. 2015 October.

[60] Slaney M. Auditory toolbox. Interval Research Corporation, Tech Rep. 1998;10:1998.

[61] Mills AW. Lateralization of high-frequency tones. The Journal of the Acoustical Society of America. 1960;32 (1):132–134.

[62] Yost WA, Dye Jr. RH. Discrimination of interaural differences of level as a function of frequency. The Journal of the Acoustical Society of America. 1988;83(5):1846–1851.

Chapter 7

Spatial Audio Signal Processing for Speech Telecommunication inside Vehicles

Amin Saremi

Abstract

Since the introduction of hands-free telephony applications and speech dialog systems in automotive industry in 1990s, microphones have been mounted in car cabins to capture, and route the driver's speech signals to the corresponding telecommunication networks. A car cabin is a noisy and reverberant environment where engine activity, structural vibrations, road bumps, and cross-talk interferences can add substantial amounts of acoustic noise to the captured speech signal. To enhance the speech signal, a variety of real-time signal enhancement methods such as acoustic echo cancelation, noise reduction, de-reverberation, and beamforming are typically applied. Moreover, the recent introduction of AI-driven online voice assistants in automotive industry has resulted in new requirements on speech signal enhancement methods to facilitate accurate speech recognition. In this chapter, we focus on spatial filtering techniques that are designed to spatially enhance signals that arrive from certain directions while attenuating signals that originate from other locations. The fundamentals of conventional beamforming and echo cancelation are explained and are accompanied by some real-world examples. Moreover, more recent techniques (namely blind source segregation, and neural-network based adaptive beamforming) are presented in the context of automotive applications. This chapter provides the readers with both fundamental and hands-on insights into the fast-growing field of automotive speech signal processing.

Keywords: automotive speech signal processing, hands-free telephony, automotive voice assistant, beamforming, acoustic echo cancelation

1. Introduction

In 1990s, first telephony systems were introduced in vehicles to enable drivers to converse in hands-free phone calls through the vehicle's embedded microphones and loudspeakers while driving [1]. To assure the audio quality during the hands-free telecommunication, a number of speech signal processing techniques are widely used. Besides of hands-free telephony, speech dialog systems have been developed to enable drivers to communicate with their vehicle functions and media contents by means of voice communication [2, 3]. In the core of a speech dialog system, there is an acoustic model that performs the speech recognition task. Speech dialog systems require high quality audio input to assure the accuracy of the speech recognition.

In a vehicle audio system, the phone is connected to the infotainment head unit via a Bluetooth communication channel which allows the driver's speech (near end) to route from the microphones mounted inside the vehicle to the other side of the tele-communication network (far end). Vice versa, the speech signal received from the far end is played on the vehicle's loudspeakers.

A major problem that typically rises in this communication system is that the far end hears a replica of their own voice back from the vehicle with a certain delay (i.e. acoustic echo). The observed acoustic echo is due to the acoustic feedback from loudspeakers to the microphones in the vehicle [4–6]. Various acoustic echo cancelation (AEC) solutions have been developed to address this issue. Most of these AEC solutions use adaptive filters that aim to simulate the acoustic path between speakers and microphones and thereby estimate and subtract the echo from the received signal [4–6].

Another major problem is that the signals captured by the microphones are contaminated with ambient noise and reverberation. The ambient noise often consists of stationary noise sources (engine noise, road noise, windows vibrations), and non-stationary cross-talks from other car occupants. To address the stationary ambient noise issues, high-pass filters were used mainly to filter out the engine noise and structural vibrational components in the captured signal [1, 2]. To address non-stationary ambient noise, adaptive algorithms have been extensively developed (e.g. [7]).

Moreover, directional microphones have been previously used in vehicles to form a spatial focus toward the driver while attenuating signals arriving from other directions. These directional microphones were often Cardioid type which were usually mounted on the ceiling of the cabin, over the head of the driver, and directed toward the driver's mouth. Most common Cardioid microphones are electret condenser components and achieve this type of directivity by means of mechanical channels mounted in their membranes. In 2000s, a new generation of microphones, known as micro electromechanical system (MEMS), were introduced to the electronic industry providing superior performance in coding the sound while having a low-cost footprint [8]. Since the mobile phone industry started to extensively deploy MEMS in their products, this type of microphones has prevailed in most telecommunication applications e.g. in tablets, wearable devices, medical systems, and automobiles [8].

A major difference between MEMS and electret microphones is that MEMS microphones, due to their specific design and miniature structure, are omnidirectional i.e. are agnostic to directions since they treat sounds arriving from all directions equally. Therefore, the desirable directivity needs to be implemented by means of external post processing. To do so, a number of MEMS microphones are placed in a certain distance from each other forming an 'array' and specific array signal processing techniques are applied to exploit the time differences and relative phase shifts across the signals captured by the microphones in the array to amplify or attenuate sounds arriving from specific directions and thereby create the desired spatial directivity [9–11].

2. Spatial filtering: beamforming

2.1 Basic concepts

A beamformer is a signal processing module that performs spatial filtering to separate signals that have overlapping temporal and frequency contents but originate from different spatial locations [9–12]. A conventional linear beamformer, as shown

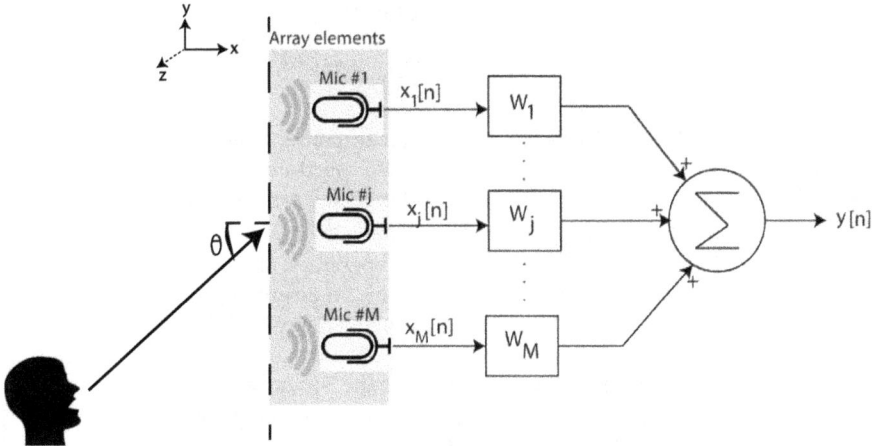

Figure 1.
A linear beamformer that consists of an array of M microphones. The signal captured by j^{th} microphone passes through the jth finite-impulse-response filter defined by its weights (W_j). The direction of arrival (DOA) is shown by θ.

in **Figure 1**, is a filter-and-sum system that consists of a number of filters that are applied on the input array signals and the results are thereafter summed. The task is to set the complex filter coefficients (W_j in **Figure 1**) in a manner to amplify specific directions in the received array signals and suppress other directions.

From another perspective, a beamformer can be viewed as a multiple-input single-output system whose output $y[k]$ is determined based on Eq. (1) below.

$$y[n] = \sum_{j=1}^{M} \sum_{p=1}^{L} W_{j,p} \, x_j[n-p] \tag{1}$$

Eq. (1) can also be viewed as a summation of M finite impulse response (FIR) filters with L coefficients per each filter that are applied on input signals ($x_j[n]$). Eq. (1) can be summarized into Eq. (2) where T denotes Hermitian (complex conjugate) transpose and W represents an $M \times L$ matrix of filter coefficients.

$$y[n] = W^T * x[n] \tag{2}$$

2.2 Conventional beamforming

A conventional beamformer assumes that each described filter needs to apply a specific tap delay (τ) on the corresponding array signal to properly align the inputs to achieve the desired directivity in the output. In this sense, each FIR filter has the following frequency response. The first filter ($p = 1$) has no associated tap delay since the signal from the first microphone is considered the zero-phase reference.

$$r(\omega) = \sum_{p=1}^{L} |W_p| \, e^{-j\omega\tau(p-1)} \tag{3}$$

Assuming that the propagating sound pressure is a complex plane wave with the direction of arrival (DOA) θ and frequency ω, the tap delay at p^{th} filter (τ_p) is a function of θ, and Eq. (3) can be re-written as below.

$$r(\omega) = W_p \, d(\omega, \theta) \tag{4}$$

$$D(\omega, \theta) = \left[1; e^{j\omega\tau_2(\theta)}; \ldots e^{j\omega\tau_M(\theta)}\right]^T \tag{5}$$

The term $D(\omega,\theta)$ is known as the array vector response. $D(\omega,\theta)$ determines the spatial outcome of the beamformer and thus is also called steering vector or direction vector. The simplest solution is to apply a constant delay per array, a so called 'delay-and-sum' algorithm. Accordingly, each array signal is delayed by $\tau_p = (p-1)\frac{d}{c}\sin(\theta)$ where c is the speed of sound, 343 m/s at 20C temperature, and p extends from 1 to M.

The distance between the array elements, d, is an essential geometric constraint that has a great effect on the performance of such a delay-and-sum configuration. An important limitation that is imposed on the performance of the beamforming due to the distance between microphones (d) is the 'spatial aliasing frequency' (f_{al}) that is calculated by $f_{al} = \frac{c}{2d}$ which gives the upper frequency limit of the delay-and-sum system. This is because, at this frequency (f_{al}), the phase shift at the microphones equals half the wavelength (λ) of the signal (see Figure 3 and Figure 4 of [12]). Therefore, to avoid spatial aliasing, the distance between the microphones (d) should be chosen carefully for the delay-and-sum beamformer to push f_{al} above the frequency range of interest.

In more sophisticated beamformers, the tap delay values (τ_p) in Eqs. (4)–(5) are set as a function of angular frequency (ω) in a *filter-and-sum* configuration. The aim is to control the behavior of the system at different frequency ranges and assure a consistent directivity across the entire frequency range of interest. A well-designed filter-and-sum beamformer with tailored frequency-dependent tap delays, $\tau_p(\omega)s$, can overcome the upper frequency barrier (f_{al}) to some good degree.

If the angles at which the interfering signals arrive is known, it is possible to design the steering vector so that the beamformer minimizes sound intensities (represented by statistical variance in the data) arriving from these specific angles. In this configuration, called linearly constrained minimum variance (LCMV) beamforming, the steering vector is designed to multiply null in given interference directions while amplifying the desired DOA.

Figure 1 presents a one-dimensional beamformer which operates in xy plane as DOA (θ) is in that plane. However, if necessary, it is possible to add microphones on z axis where similar equations, Eqs. (1)–(5), can be written in the xz plane with a DOA in that plane. Accordingly, a two-dimensional beamformer would be created which filters the xyz space with regard to one DOA in xy plane and another DOA in xz plane.

From another perspective, **Figure 1** depicts a 'broadside' beamformer which is designed to form a beam toward the target which is located in the broadside plane of the microphone array. However, if the target is located along with the axis of the array (therefore $\theta = \pm90$), then the configuration is called 'end-fire' [9]. In an end-fire configuration, the summation in **Figure 1** is replaced by subtraction. Consecutively, each filter output is subtracted in Eq. (1) instead. Thus, an end-fire beamformer is also called a 'filter-and-subtract' or a 'differential' beamformer. This type of beamformer, which can be viewed as a special case of the general beamforming shown in **Figure 1**, forms a beam toward either above the array axis ($\theta = 90$) or below the array axis ($\theta = -90$).

2.2.1 Fixed beamforming vs. adaptive beamforming

In fixed beamforming, the DOA is known and time-invariant thus the steering vector, $D(\omega,\theta)$, can be set for a known fixed geometry. A good example of fixed beamforming is in automotive industry where the target talker (driver) sits in a fixed location and the DOA toward the microphones is predetermined. Fixed beamforming can be viewed as a 'data-independent' algorithm since the steering vector is designed solely based on the known geometry of the sound propagation and is independent of the received data. In contrast, in adaptive beamforming, DOA varies and the steering vector should adapt to the changes in DOA. For an example, an adaptive beamformer is needed in case the system is supposed to localize and adapt itself to capture signals from all car occupants (besides of the driver) who are sitting at different location inside the vehicle. In this case, the system should itera-tively find the target talker first and then update its steering vector toward that target. Another example of an adaptive beamformer is in the 'cocktail party' prob-lem wherein the target location can vary in the room and the system should con-stantly localize the target and the beamforming algorithm should adapt to the new DOA and other geometrical factors, accordingly. From this perspective, adaptive beamformers can be viewed as being 'data-dependent' systems since their parame-ters change according to variations in the received data. As a result, adaptive beamformers usually require substantial computational resources [10, 13, 14].

An adaptive beamformer is often accompanied by a pre-processing stage whose task is to localize the target and determine the new DOA. This 'localization' stage usually accomplishes its task by examining the data and finding optimum DOA that maximizes a specific metric such as signal strength, or speech intelligibility [10, 13–15]. Alternatively, some localization algorithms are built on minimizing a specific cost function, such as noise and reverberation, in the signal. When the localization algorithms finds the DOA, the values in the steering vector $(D(\omega,\theta))$ should adapt to this new angle.

There are some relatively newer solutions that merge the 'localization' and 'beamforming' stages together. Warsitz and Haeb-Umbach [14] presented an algo-rithm that optimizes the FIR filter coefficients (denoted by W in Eqs. (1)–(2) above) by iteratively estimating and maximizing the cross power spectral density of the microphone signals. An important feature of this algorithm is that the filter coeffi-cients are optimized directly without localizing the source. In other words, the DOA information is implicitly absorbed in the optimization problem although it is possible to extract the underlying DOA information from the results afterwards, if needed.

2.3 Neural-based adaptive beamforming in speech recognition applications

Speech signal enhancement (SSE) techniques, such as beamforming, have tradi-tionally been performed as an independent pre-processing stage to speech recogni-tion back ends [13, 15]. In this conventional setup, SSE algorithms are performed to improve the signal-to-noise ratio (SNR) by reducing ambient noise and reverbera-tion in the captured signal. The output of the SSE stage is then fed into acoustic models, usually deep neural networks, which perform the automatic speech recog-nition (ASR) task.

In the last few years, adaptive beamforming algorithms have been designed that are tuned jointly together with the speech recognition backend [13, 15, 16]. To do so, the FIR coefficients (shown by W in **Figure 1** and also in Eqs. (1)–(2)) are jointly trained together with the parameters of the ASR model where the optimization is performed using a gradient learning algorithm. The goal of this optimization pro-cess is to find FIR coefficients that result in higher ASR accuracy.

Several neural-network approaches have been developed to address the ASR problem [15] but the most successful ASR models are currently built on the convolutional deep neural network (CL-DNN) concept [13, 15]. The input is filtered by a time-domain filterbank pre-processor, usually a Gammatone filterbank together with a nonlinearity function, which is supposed to loosely mimic the human auditory periphery (cochlea) in terms of spectral feature extraction and compression [17]. The output is then fed into the CL-DNN model. The first stage in the CL-DNN model is the *fconv* layer that convolves the output signals across the filterbank channels and the results are pooled along the frequency axis. The next stage comprises a number of long short-term memory (LSTM) layers. LSTM network is a specific type of recurrent neural network that is tailored for recognizing sequential time-series data such as audio. The final stage is a single fully-connected DNN that consists of at least 1024 hidden units [13, 15, 16].

Sainath et al. [13, 16] presented a multi-microphone solution to incorporate the data captured by M microphone arrays into the CL-DNN model. They replaced each spectral channel of the Gammatone filterbank pre-processor with FIR filters that are connected to the microphones and are used for beamforming (identical to Eqs. (1) and (2) above). They essentially created a filter-and-sum beamformer per spectral channel. The difference is that the tap delays (T_p) and therefore DOA data are implicitly absorbed in the FIR coefficients similar to earlier works by Warsiz and Haeb-Umbach [10]. Sainath et al. [13, 16] trained the beamforming FIR coefficients together with the CL-DNN parameters using a gradient learning algorithm to maximize the ASR accuracy. Sainath et al. [16] showed that during the training, the FIR coefficients become optimized to extract both spectral and spatial features of the incoming speech signals. They showed that the multi-microphone ASR model with joint beamforming achieves an over 10% improvement in word error rate (WER) compared to its single-microphone counterpart.

Besides of excellent ASR accuracy, a major benefit of neural-network based beamforming is that the model is, to a great extent, independent of the array spacing whereas the conventional beamforming relies on the prior knowledge of the distance between microphones (d) to calculate the tap delays. Due to its remarkable success in ASR, neural-based beamforming is prevailing in all ASR systems that have access to multiple microphone input. A very good candidate for applying this technique is in automotive ASR systems wherein online voice assistants based on this technique are currently being designed and evaluated.

A potential shortcoming of the neural-based beamforming is that the source localization information (i.e. DOA) is implicitly embedded in the model and might not be extractable and interpretable in terms of physical geometry. This could impose a limitation in applications which require an explicit knowledge of the source location. Besides, an important distinction is that neural-based beamforming parameters are tuned solely based on ASR objectives and might not necessarily improve the audio quality (e.g. SNR) with regard to the human psychoacoustics [13, 15, 16]. Therefore, neural-network based beamforming is currently considered more applicable to speech recognition tasks rather than to applications such as telephony wherein human listeners are involved. The feasibility of neural-network based beamforming for telephony applications and its relation to human psychoacoustics need further investigations.

2.4 Beamforming applications in automotive industry

Beamforming techniques introduced into the automotive industry almost at the same time that the first automotive hands-free telephony and speech dialog systems were being devised [1]. Although there have been some studies using multiple microphones [18], it is by far more common to only have dual microphones available in vehicles for beamforming. There are mainly two reasons for this, the first one is the production costs, and the second one complications in the vehicle's interior design and excess wiring if multiple microphones are used. Therefore, in the following sections regarding automotive applications, two-microphone solutions are in focus.

Figure 2(A) shows a car with two dedicated microphones (marked M1 and M2) about 4.5 centimeters apart (d = 4.8 cm) that have been mounted in the car ceiling. The DOA is ideally around 90 degrees according to the illustrated coordinates. To provide a fixed beamforming solution for this particular geometry, an 'end-fire' differential beamformer should be used since the desired source is located along the axis of the array (θ = $\pm$90). The input signals are filtered according to Eqs. (1)-(5) and then subtracted. The frequency-dependent tap delays for microphone M2 (i.e., $\tau_2(\omega)$) were chosen to enable the steering vector to enhance sounds that arrive from the driver side (θ = 90).

Figure 2 shows the beam patterns resulted from an end-fire filter-and-sum beamformer where the tap delays for M2 microphone (i.e. $\tau_2(\omega)$) have been adjusted as a function of frequency at several frequency channels covering the frequency range from 0.1 to about 7 kHz. **Figure 2(B)** shows the beam pattern at 1 kHz. This beam pattern demonstrates that sounds from θ = 90 (driver side) have passed through the system whereas sounds from θ = −90 = 270 have been substantially attenuated. A very similar beam pattern is shown by **Figure 2(C)** at 2 kHz

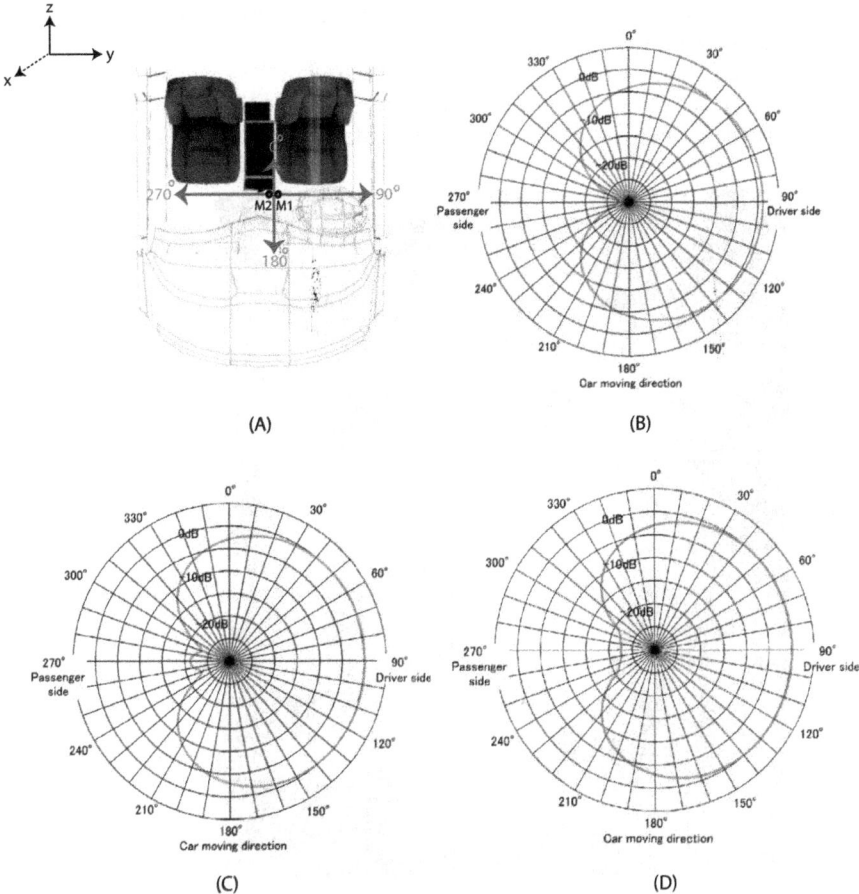

Figure 2.
A) a car cabin geometry with a dual microphone mounted in the car ceiling. B) the beampattern achieved by the described end-fire (differential) beamformer at 1 kHz, C) at 2 kHz, and D) at 4 kHz. (A) dual microphones in a personal car, (B) beampattern at 1KHz, (C) beampattern at 2KHz and (D) beampattern at 4KHz.

although the beam pattern at 4 kHz, shown in **Figure 2(D)**, deviates somewhat with minimal effect on overall performance.

The presented beamformer was tested in-situ by placing a head-and-mouth simulator system at the typical location of the driver's head and playing standard hearing-in-noise test (HINT) sentences [19] while the engine was running in idle mode creating some stationary background noise. The raw signals captured by the M1 microphone were recorded. The test was repeated while applying the described beamforming on the raw signals. The beamforming results were compared to the raw signal. The results showed a signal-to-noise ratio improvement (SNRI) of 5.7 dB (A) across frequencies between 0.1 and 8 kHz.

Figure 3 shows the beamforming geometry in a large truck cabin wherein a dual microphone array is installed on the overhead compartment. The distance between the array and the driver's mouth is about 0.4 m and the DOA is approximately 30 degrees ($\theta = 30$ in zy plane) although these numbers vary depending on the height and other biometrics of the driver. The distance between the two microphones (d) is 23 mm which yields a higher spatial aliasing frequency upper limit compared to the system shown in **Figure 2**. In this case, an end-fire beamformer can be designed to form a beam downwards toward the cabin's floor ($\theta = 90$). The drawback is that an amount of engine noise and AC fan noise will also leak into the beamformer since these noise signals originate from the dashboard which is also located below the overhead compartment.

Alternatively, a broadside beamformer can be used to direct the beam toward the DOA of $\theta = 30$. As a well-known drawback, the broadside configuration also amplifies the angle that is 180 degrees behind the DOA (i.e. 30 + 180 = 210

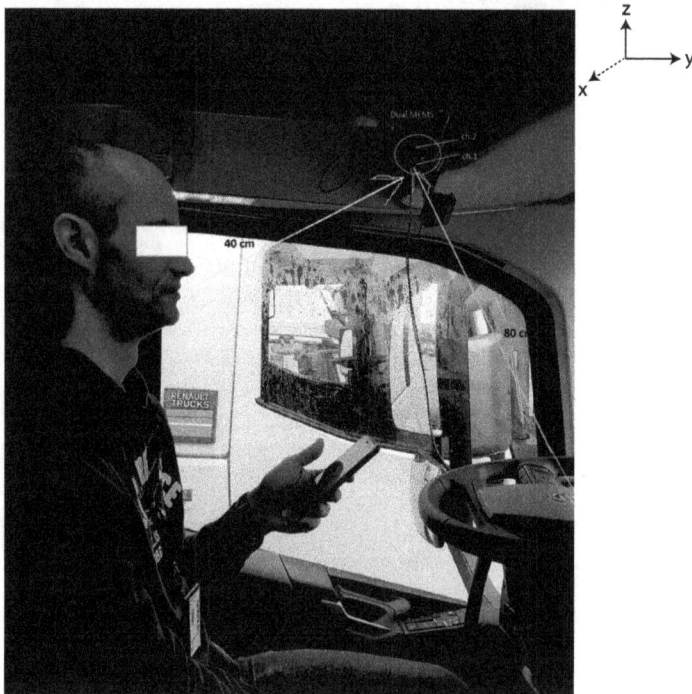

Figure 3.
A truck cabin geometry with a dual microphone installed in an overhead compartment forming DOA of approximately 30 degrees ($\theta = 30$) toward the mouth of a 180-cm long male driver. The yellow arrow shows the first-wave propagation from the driver's mouth whereas the green arrow shows the noise signal propagation (engine noise and AC fan noise).

degrees in this case). This is because the broadside beamforming, characterized by Eqs. (1–4), is agnostic to the 180-degrees axis and any sound coming from $\theta + 180$ is treated equally as θ. However, since the overhead console behaves as a mechanical damper for sounds and vibrations coming from the roof and the backward direction, broadside solution appears to be a better solution in this practical case.

A broadside filter-and-sum has been devised with tailored frequency-dependent tap delays to facilitate a consistent beamforming toward the driver at frequencies between 0.1 and 8 kHz. The in-situ measurements and beam patterns are not finalized yet. However, the preliminary analysis indicates that the system can achieve an SNRI of about 6 dB when the engine is running in idle mode. The described beamformer functions on xz plane. However, a third microphone can be added on y axis next to the existing pair of microphones to perform beamforming on xy plane as well. This new beamformer on xy plane can be tuned to attenuate sounds arriving from the co-driver's side although adding multiple microphones are currently uncommon in vehicles.

3. Acoustic echo cancelation

3.1 Basic concepts

Acoustic echoes are generated in speech telecommunication networks due to the acoustic feedback from loudspeakers to microphones. This phenomenon deteriorates the perception of sound by causing the users to hear a delayed replica of their own voice being reflected back from the other side of the network [4–6]. **Figure 4** shows a driver in a truck cabin making a phone conversation through embedded microphones and loudspeakers (marked red). The speech signal is denoted by $s[n]$ whereas the echo is denoted by $y[n]$ and $r[n]$ represents the ambient noise. The echo ($y[n]$) can be considered a copy of the far-end speech signal played by the loudspeaker ($x[n]$) that has been filtered by the acoustic path (modeled by an FIR filter with the linear impulse response $h[n]$) between the loudspeaker and the microphone. The received signal ($d[n]$) is an addition of these three signals ($d[n] = s[n] + y[n] + r[n]$).

To remove acoustic echoes from the captured signal, acoustic echo cancelation (AEC) algorithms have been developed that use machine learning methods to adaptively estimate the 'acoustic echo path' in real time and subtract its effect from the captured signal so that only the desired near-end speech components remain [4–6, 20, 21]. Similar adaptive methods are also commonly used for estimating the noise propagation acoustic path in stationary noise reduction applications (e.g. [7]). The most common adaptive method used in AEC tasks is normalized least mean square (NLMS) filtering [5, 6, 20, 21]. Least-mean-square adaptive filters were used even in the earliest generation of AECs [4] and several improved variants of it, such as NLMS, have been developed since then [5, 6, 20, 21].

The true impulse response of the echo path (i.e. $h[n]$) is unknown and the task of an AEC solution is to identify it. To do so, the NLMS algorithm constantly tries to adapt to an impulse response ($\hat{h}[n]$) that closely matches the true impulse response of the echo path (i.e. $\hat{h}[n] = h[n]$) and consequently, $\hat{y}[n]=y[n]$ and thus the error signal, $e[n]$, becomes zero. The length of $\hat{h}[n]$, denoted by L, has an important role in the performance of the AEC. The filter should be long enough to realistically model the acoustic path and furthermore to guarantee that the acoustic path can be assumed time-invariant during the time that corresponds to L samples. If the goal of

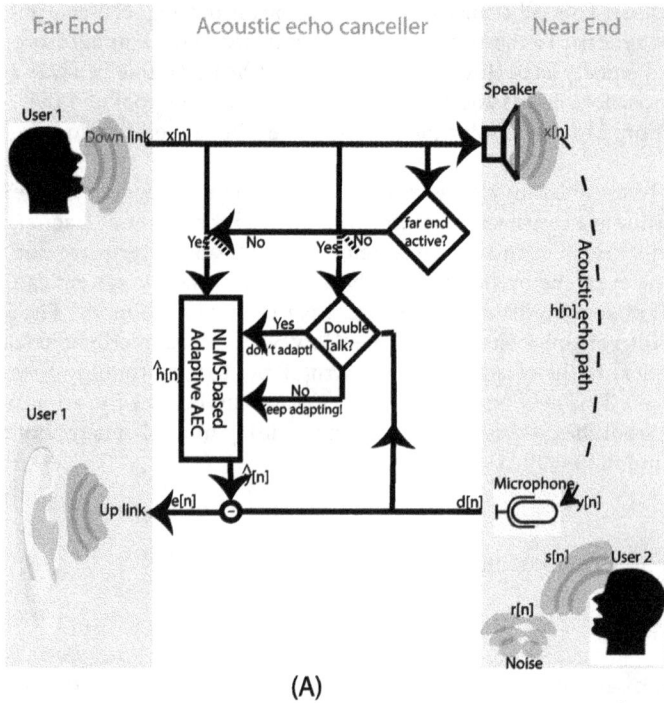

(A)

(B)

Figure 4.
A) an overview of the described AEC algorithm. B) a driver in a truck serving as the near-end party during a hands-free phone conversation.

the AEC is to reduce the echo by 30 dB, then L should correspond to T30 reverberation time [5]. In most modern vehicles, 50 ms appears to be a good estimate of T30.

In case the echo ($x[n]$) is effectively the only signal present ($r[n]$ and $s[n]$ are absent), the output of the adaptive process ($\hat{y}[n]$) is given by Eq. (6) below where L tap of $x[n]$ is transposed (represented by $x_L^T[n]$) and multiplied by $\hat{h}[n]$.

$$d[n] = \hat{y}[n] = x_L^T[n].\hat{h}[n] \tag{6}$$

The adaptive process estimates a new $\hat{h}[n]$ per each sample of y[n] through a small adjustment $\Delta\hat{h}[n]$ in each iteration, as expressed in Eq. (7). This adjustment value is determined based on the error signal and the reference signal according to Eq. (8) where $\mu[n]$ is known as 'step size'. Choosing an optimized step size is important for the convergence rate and accuracy of the system and is determined by the parameters α and β. These two parameters need to be adjusted according to the specifics of every given NLMS problem. Choosing appropriate values for α and β has been comprehensively studied to a great complication [5, 20–22].

$$\hat{h}[n+1] = \hat{h}[n] + \Delta\hat{h}[n] \tag{7}$$

$$\Delta\hat{h}[n] = \mu[n] \times (x_L[n] \times e[n]) \tag{8}$$

$$\mu[n] = \frac{\alpha}{\beta + \delta^2_{x_L[n]}}, \delta^2_{x_L[n]} = var(x_L[n]) = x_L{}^T[n].x_L[n]. \tag{9}$$

Eqs. (6)–(9) are applicable only when the echo is the only present component in the received signal (i.e. d[n] = y[n]). In other words, the near-end talker is silent and the ambient noise is insignificant (s[n] = 0 and r[n] 0). However, in a natural full-duplex speech communication both parties (far end and near end) might talk simultaneously sometimes (i.e. a 'double-talk' event may occur). If there is any remarkable double talk in d[n] (i.e. a non-zero s[n]), then the adaptive process, formulated by Eqs. (7)–(9), might diverge and fail since the s[n] components cannot be modeled by h[n]. Therefore, every adaptive AEC solution needs to constantly watch out for double-talk events to halt the adaptation as long as double talk is present [23, 24].

3.2 Spatial acoustic echo cancelation

All conventional NLMS-based adaptive methods, explained in the previous section, rely on modeling the acoustic path by an FIR system and aim to find the coefficients of the corresponding filter (i.e. h[n] in **Figure 4**). A major drawback of NLMS-based adaptive approach is that the adaptive process, presented by Eqs. (7)–(9), needs to run constantly which imposes a remarkable computational cost. This is because the acoustic path, characterized by h[n], constantly changes due to slight movements of the objects in the environment and other reasons such as temperature variations and the adaptive process needs to estimate the new impulse response.

Recently, alternative methods based on probabilistic clustering techniques have been successfully used for blind source separation (BSS) of the echo components from the near-end speech signal [25]. The BSS method uses the spatial information from the captured microphone signals to cluster and separate the desired speech signal (s[n]) from the echo (y[n]). Any BSS method, similar to beamforming, requires multiple microphones to be able to extract the location cues.

Every BSS method assumes that the signal captured by the microphones $(d_1, d_2, \ldots, d_M)$, where M denotes the number of microphones, are from N independent source signals $(s_1, s_2, \ldots, s_N)$ that have been mixed together. The mixture model is then modeled as described by Eq. (10) below where h_{jk} is the impulse response of length L that describes the acoustic path from the kth source (S_k) to the jth microphone (d_j).

$$d_j[n] = \sum_{k=1}^{N}\sum_{p=1}^{L} h_{jk}(p) * s_k[n-p] \tag{10}$$

$$d_1[n] = \sum_{p=1}^{L} h_{1s}(p) * s[n-p] + \sum_{p=1}^{L} h_{1x}(p) * x[n-p] \qquad (11)$$

$$d_2[n] = \sum_{p=1}^{L} h_{2s}(p) * s[n-p] + \sum_{p=1}^{L} h_{2x}(p) * x[n-p] \qquad (12)$$

$$d[n] = [h_{1s}, h_{1x}; h_{2s}, h_{2x}] * [s[n]; x[n]] = W * [s[n]; x[n]] \qquad (13)$$

The BSS techniques that are extensively used to address the 'cocktail party' problem aim to find the mixing impulse responses (h_{jk}) and use this information to de-mix and find the original speech signals [25, 26]. In case there are more microphones than sources ($M \geq N$), the BSS reduces to a 'determined' problem and linear filters can successfully be deployed to effectively separate the mixtures. Otherwise, if there are fewer number of microphones than sources ($M < N$), then the problem is 'underdetermined' and linear filters would not work adequately.

In the case depicted by **Figure 3**, there are two microphones and also two independent sources ($M = N = 2$), namely: 1) near-end speech ($s[n]$), and 2) the echo ($x[n]$) that leaks from the loudspeaker to the microphones. Eqs. (11)–(12) formulize the mixture model for the signal received by microphone 1 and microphone 2, respectively. Here, h_{1x} and h_{2x} are the impulse responses of the acoustic path from the loudspeaker to the first and the second microphone, respectively. Eqs. (11)–(12) can be summarized into a matrix form and re-written by Eq. (13) where the relation between the sources and microphones signals is denoted by a Wiener filter. Eq. (13) can be used to inverted so that $[s[n]; x[n]] = W^T * d[n]$.

A conventional approach to solving Eqs. (10)–(13) is using independent component analysis (ICA) [26]. Accordingly, a cost function is defined to estimate the statistical (convolutional) independence of $s[n]$ and $x[n]$. The coefficients of W (which comprises h_{1x}, h_{1s}, h_{2x}, and h_{2s}) are adaptively updated so that the statistical independence of $s[n]$ and $x[n]$ is increased. The statistical independence is often increased by either maximizing the non-Gaussianity or by minimizing the mutual information between the two signals.

3.3 The performance of acoustic echo cancelers in vehicles

The performance of an AEC is primarily measured by two metrics: 1) echo return loss enhancement (ERLE), and 2) convergence time. ERLE is a commonly used indicator for quantifying the achievement of an AEC solution to attenuate echoes [5, 20, 21, 23]. ERLE is calculated according to the Eq. (14) below where $\sigma^2_{d[n]}$ and $\sigma^2_{e[n]}$ represent the variance of the captured audio by the microphone ($d[n]$) and the variance of the error signal ($e[n]$) which is the output of the AEC and is ideally echo-free. Since all signals are zero-mean, the variance of a signal is a measure of the magnitude of its intensity. Therefore, Eq. (14) yields the ERLE as the magnitude of the AEC output relative to the microphone input signal.

$$ERLE = 10 \times log_{10} \frac{\sigma^2_{d[n]}}{\sigma^2_{e[n]}} \qquad (14)$$

International telecommunication union (ITU) G.168 standard for AECs [27] declares a number of requirements that should be followed in all speech telecommunication applications. Accordingly, the AEC should yield at least 6 dB of ERLE at the second frame (since each frame is 50 ms in a typical automotive solution, this means at 0.1 second). The ERLE should then increase to minimum 20 dB at 1 second. Thereafter, the ERLE should reach its steady state at 10 second and should stay over that steady state value, afterwards.

The convergence time is the time it takes for the AEC to reach to its steady-state ERLE. ITU G.168 requires that the convergence time should be no longer than one second. In the tuning of the adaptive parameters, such as step size, there is a tradeoff between ERLE and convergence time since higher ERLE might result in slower convergence time [21, 22].

We implemented an adaptive NLMS-based AEC described by Eqs. (6)–(9) on a large Volvo truck model. The length of the Wiener filter (L) was chosen 800 which corresponds to 50 ms at the sampling rate of 16 kHz which would be consistent with T30 in large vehicles. The term α in Eq. (9), which could take a value between 0 and 2, determines the speed of convergence. Higher α values result in quicker adaptation of the NLMS algorithm, however, there is a tradeoff between convergence and overall success of the echo canceller in terms of ERLE ([20]). Here, we chose $\alpha = 1.98$ to assure the fast convergence of the algorithm. The term β, known as the regularization parameter, is meant to improve the performance of the NLMS in noise and it has to be adjusted with regard to the characteristics of the ambient noise ($r[n]$ in **Figure 1**) and the signal-to-noise ratio (SNR) of the microphone hardware [20]. Here, we chose $\beta = 0.1$ which corresponds to the SNR of the electret condenser microphones that are commonly used in automotive industry.

Furthermore, a statistical double-talk detection (DTD) decision circuit based on the normalized cross-correlation (NCC) between $x[n]$ and $d[n]$. NCC is also called 'Pearson correlation coefficient' in statistics [28]. In case the far-end is the only talker, there will be a non-zero cross-correlation between $x[n]$ and $d[n]$. However, when the near end talks too (i.e. DT occurs), the cross-correlation between $x[n]$ and $d[n]$ diminishes and approaches zero since $d[n]$ would convey $s[n]$ components as well. Accordingly, DT is detected if NCC drops below a certain threshold. Eq. (15) presents the NCC between $x[n]$ and $d[n]$ where $\sigma_{x_L[n]}$ and $\sigma_{d_L[n]}$ are the standard deviation (square root of variance) of L samples of $x[n]$ and $d[n]$, respectively, and $cov(x_L[n], d_L[n])$ is the covariance between them.

$$NCC\,(x_L[n], d_L[n]) = \frac{cov(x_L[n], d_L[n])}{\sigma_{x_L[n]} \times \sigma_{d_L[n]}} \tag{15}$$

NCC can yield a number in the range [−1, +1], where +1 indicates perfect correlation and − 1 perfect anti-correlation between the two inputs while 0 shows a non-existing correlation. Here, we set the threshold of our DTD decision to 10^{-4} using the method discussed in [28] by normalizing the false alarm probability (pf) to about 0.1.

To evaluate the presented AEC solution, the far-end party reads 10 HINT sentences while the driver (near-end party) is silent and the vehicle's engine is off. The system registers the incoming signal to the speaker ($x[n]$) while the microphone records $y[n]$. In this case $y[n] = d[n]$ since the driver is silent ($s[n] = 0$) and there is no engine noise ($r[n] = 0$). The presented solution is applied on these signals and, as depicted in **Figure 5** below, the presented AEC solution manages to attenuate the echo received by the microphone significantly by a total of 25.54 dB according to Eqs. (10)–(13). **Figure 5(B)** shows ERLEs per each sentence and how the ERLE becomes stronger as the algorithm continues adapting. The results demonstrate compliance with ITU G.168 standards [27].

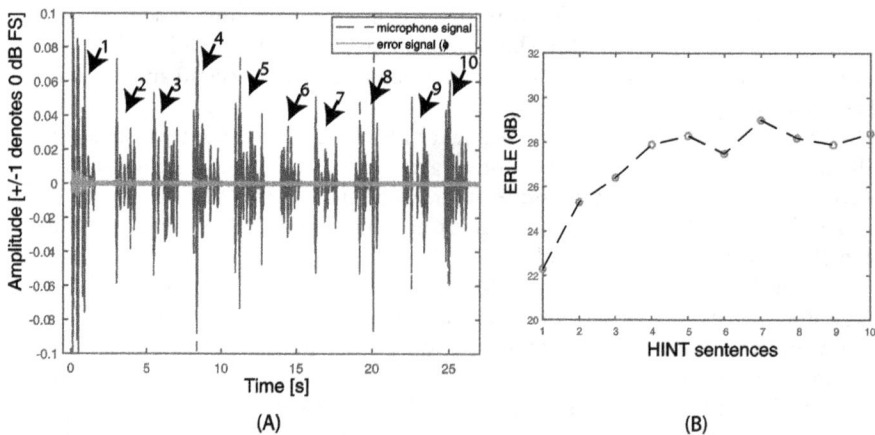

Figure 5.
A) Captured microphone data (d[n]) versus the output of the AEC (e[n]) while the far end is reading ten HINT sentences. The sentences are marked by numerical indicators. B) the echo attenuation achieved by the presented AEC solution in terms of ERLE per HINT sentence.

3.4 Post-processing acoustic echo suppression

The minimum acceptable ERLE required by ITU G.168 (i.e. 20 dB) may not practically suffice since the echo might still be noticeable and irritating especially if the loudspeaker volume is set at a high level. As an example if the loudspeaker is set to generate sounds that are about 70 dB SPL loud, an ERLE of 20 dB would imply that there is an echo of 50 dB SPL (i.e. 70–20 = 50) being transmitted back to the far-end party which can be quite noticeable. Therefore, it is good practice in automotive industry to achieve much higher echo reduction i.e. typically over 40 dB.

The conventional NLMS-based adaptive AEC modules typically achieve maximum 30 dB ERLE, as shown in **Figure 5**. Therefore, to further improve the echo reduction, the remaining echo components (i.e. 'residual echo') are suppressed by means of acoustic echo suppression (AES) post processing. The simplest AES methods which have historically been used are based on attenuating the captured microphone signal (*d[n]* in **Figure 4**) whenever the farend is talking (i.e. whenever the magnitude of *x[n]* is over a reasonable threshold) [5]. A major shortcoming of this method is that, in case of double talk wherein both near end and far end are simultaneously talking, the near-end speech signal is also attenuated. Another issue is that such an approach is nonlinear. Speech recognition models require that the audio signal chain must be free of any nonlinearity [29]. Since adaptive AEC algorithms use linear filters to cancel echo, they could legitimately be used as a pre-processing stage to ASR systems. However, the use of nonlinear AES must be avoided in ASR applications. As a result, linear solutions, such as BSS techniques explained previously by Eqs. (10)–(13), have been deployed to perform the task of AES on the residual echo especially in speech recognition applications. A properly designed combination of conventional adaptive AEC and a post-processing AES must comfortably achieve echo reductions over 40 dB.

4. Discussions, conclusions, and prospects

Hands-free telephony has been extensively offered in premium cars since early 2000s, and since then, audio signal processing modules have been deployed to enhance the speech signal quality by means of addressing issues such as acoustic echo, ambient noise, and reverberation [1]. Besides of hands-free telephony, speech dialog systems have been developed to enable drivers to communicate with vehicle functions by means of voice communication [2, 3]. In the core of such a speech dialog system, there is a neural-network based acoustic model that performs the speech recognition task. Speech recognition systems also demand high quality audio input which makes speech signal enhancement techniques necessary. Especially, online voice assistants rely on specific 'wake words' (also called 'hot words') to communicate with users. These are 'Ok Google!' for Google assistant, 'Alexa!' for Amazon Alexa, and 'Hey Siri!' for Siri. The ASR system should constantly listen for these wake words meanwhile music or speech signals might be simultaneously playing on the speakers. In order to detect the wake words while playing sounds, the system needs to benefit from a capable echo cancelation module to estimate and cancel the feedback from speaker(s) to the microphone(s) as well as a noise reduction module (such as beamforming) to minimize the reverberation and ambient noise in the captured signals.

In this chapter, the fundamentals of the filter-and-sum beamforming were described and two practical designs of dual-microphone fixed beamforming (end-fire versus broadside) were presented inside a personal car and a truck, respectively. The fundamentals of beamforming were described for a general case although the applications were exclusively focused on dual microphones because that is the most common setup in vehicles. The directivity index, which is the gain of the beamforming on the desired DOA relative to all other directions, is a good measure of a beamformer's performance. A conventional multi-microphone fixed beamformer can achieve a directivity index of about 25 dB at best [30]. In real world, the directivity index turns out to be lower. In case of dual-microphone solution, the directivity index is minimal i.e. in the range of 10 to 12 dB. Multiple microphones can provide a sharper beam and potentially higher SNRIs.

Despite its modest directivity index, a well-designed beamforming system improves the quality of the sound substantially. One important benefit of beamforming, besides of the SNRI, is the reduction in the perceived reverberation. Reverberation is related to the sum of all sound reflections from the walls and surroundings of a given acoustic room and has been shown to have adverse effects on the speech intelligibility especially in case of hearing-impaired listeners [31]. Beamforming minimizes reverberation in the captured signal by means of geometrically dampening the sound reflections received from undesired directions and thereby facilitates speech intelligibility. Moreover, beamforming modules are in many cases followed by non-stationary noise reduction modules that adaptively suppress the noise (e.g. [7]). Together with the beamformer, an adaptive noise suppressor can achieve very good results in managing non-stationary noise.

Neural-based beamforming was also described in this chapter. This type of beamforming, wherein the steering filter coefficients are optimized jointly together with a neural-network speech model, has emerged in many speech recognition applications and shown remarkable success [13, 15, 16]. However, since the beamforming coefficients are optimized implicitly as a part of a speech recognition task, the success of this method in improving sound quality for a human listener is not entirely known and further studies are needed to evaluate this method for telephony and hearing-aid applications wherein human listeners are involved.

A large part of this chapter was dedicated to acoustic echo cancelation. The fundamentals of a conventional adaptive method based on NLMS was described. In this method, the acoustic path between the loudspeaker and microphone is modeled by an FIR filter and the adaptive process seeks to find the coefficients of this filter and subtract the echo from the captured signal. Adaptive NLMS-based acoustic echo cancelers are relatively easy to implement and are extensively in use. If designed appropriately, this method can comfortably achieve ERLEs about 30 dB [5, 21, 22, 30]. Although this level is higher than the required level by the ITU guidelines [27], a higher ERLE becomes necessary in most automotive telephony applications. Therefore, acoustic echo suppression algorithms have been developed as post-processing modules to further reduce the residual echo.

The simplest and most common acoustic echo suppressors are implemented by means of applying a gain on the microphone signal and reducing this gain whenever the far-end party is talking. However, due to its nonlinear behavior, this approach cannot be used in speech recognition applications which require linearity of all audio components [29]. Instead, linear approaches such as BSS based on ICA appear to be suitable. The BSS method uses spatial cues to find mixing coefficients of a linear model and uses this information to de-mix the signals and segregate the source signals (in this case: echo versus near-end speech).

Although beamforming and echo cancelation are well-known problems that have been extensively studied since early 1960s [4, 5, 9, 30], it needs great efforts to tailor them to address new challenges. Therefore, new statistical optimization approaches and neural-network based solutions are being deployed to strengthen the conventional methods, whenever feasible. Automotive industry is expanding quickly and manufacturers are competing in providing vehicles that allow vehicle occupants to have independent conference calls simultaneously. Another competition frontier is speech recognition. Automotive manufacturers aim to provide user interfaces that are driven by voice. These interfaces allow the drivers to simply talk to their cars and do their daily errands (such online shopping, scheduling meetings, listening to audio books) while driving by solely voice commands. Prototypes of such online automotive voice assistants have just been introduced as Google [32] and Amazon entered the game [33] and have received a great attention from the media, and the public. These systems open up new scientific and technical challenges in human-machine interfacing, cloud-based and embedded speech recognition, and last but not least, spatial audio signal processing.

Acknowledgements

Parts of this project has been funded by the innovation office at the Department of vehicle connectivity (VeCon) at Volvo group in Gothenburg, Sweden and some of the results were published in an M.Sc. thesis by Balaji Ramkumar in collaboration with Linköping University.

Author details

Amin Saremi
Department of Applied Physics and Electronics, Umeå University, Umeå, Sweden

*Address all correspondence to: amin.saremi@uni-oldenburg.de

IntechOpen

References

[1] Oh S, Viswanathau V, Papamichalis P. Hands-free voice codcation in an automobile with a microphone array. PTOC ICASSP. 1992:281-284

[2] Heisterkamp P. Linguatronic-product-level speech system for Mercedes-Benz cars. In proceedings of the first international conference on human language technology research. USA; 2001

[3] Chen F, Jonsson IM, Villing J, Larsson S. Application of speech technology in vehicles. In: Speech Technology: Theory and Applications. UK: Springer; 2010. pp. 195-219

[4] Sondhi MM, Presti AJ. A self-adapting echo canceller. Bell System Technical Journal. 1966;45:1851-1854

[5] Kellermann W. "Echo Cancellation,"in Handbook of Signal Processing in Acoustics. Vol. 1. USA: Springer; 2008. pp. 883-895

[6] Jung MA, Elshamy S, Finscheidt T. An automotive wideband stereo acoustic echo canceler using frequency-domain adaptive filtering. 22nd Europen signal processing conference (EUSIPCO). 2014. pp. 1453-1456

[7] Chen YH, Raun SJ, and Qi T. An automotive application of real-time adaptive wiener filter for non-stationary noise cancellation in a car environment. IEEE international conference on signal processing, communication, and computing (ICSPCC). 2012. pp. 597-601

[8] Zawawi SA, Hamzah AA, Majlis BY, Mohd-Yasin F. A review of MEMS capacitive microphones. Micromachines. 2020;11(482):1-28

[9] Van Veen BD, Buckley KM. Beamforming: A versatile approach to spatial filtering. IEEE ASSP MAGAZINE. 1989:740-761

[10] Timofeev S, Bahai ARS, Varayia P. Adaptive acoustic beamformer with source tracking capabilities. IEEE Transactions on Signal Processing. 2008;56(7):2812-2819

[11] Vu NV, Ye H, Wittington J, Delvin J, and Mason M. Small footprint implementation of dual-microphone delay-and-sum beamforming for in-car speech enhancement. IEEE international conference on acoustics, speech, and signal processing. 2010. pp. 1482-1485

[12] Cigada A, Lurati M, Ripamonti F, Vanali M. Beamforming method: Supression of spatial alliasing using miving arrays. Journal of acousticsl Society of America (JASA). 2008;124(6):3648-3658

[13] Sainath TN, Weiss RJ, Wilsom KW, Naraayanan A, Bachiani M, Senior A. Speaker localization and microphone spacing invariant acoustic modeling from raw multichannel waveforms. Google Research. 2015:1-7

[14] Wartsiz E, Haeb-Umbach R. Acoustic filter-and-sum beamforming by adaptive principal analysis. ICASSP. 2005:797-800

[15] Hinton G, Deng L, Yu D, Dahl GE, Mohamed A-r, Jaitly N, et al. Deep neural networks for acoustic modeling in speech recognition: The shared views of four research groups. Signal Processing Magazine, IEEE. 2012;29(6):82-97

[16] Saniath TN, Weiss RJ, Wilson KW, Li B, Narayanan A, Variani E, et al. Multichannel signal processing with deep neural networks for automatic speech recognition. Google Research. 2017:1-14

[17] Saremi A, Beutelmann R, Dietz M, Ashida G, Kretzberg J, Verhulst S. A comparitive study of seven human cochlear filter models. The Journal of

the Acoustical Society of America. 2016; **140**(3):1618-1634

[18] Qi Z, Moir TJ. Automotive 3-microphone noise canceller in a frequently moving noise source environment. International Journal of Information and Communication Engineering. 2007;**3**(4):297-304

[19] Hällgren M, Larsby B, Arlinger S. A Swedish version of the hearing In noise test (HINT) for measurement of speech recognition. International Journal of Audiology. 2006;**45**:227-237

[20] Paleologu C, Ciochin S, Benesty J, Grant SL. An overview on optimized NLMS algorithms for acoustic echo cancellation. EURASIP Journal on advances in signal proc. 2015. DOI: 10.1186/s13634-015-0283-1

[21] Enzner G, Buchner H, Favrot A, Keuch F. Acoustic echo control. In: Academic Press Library in Signal Processing. USA: Academic Press; 2014. pp. 807-877

[22] Hänsler E, Schmidt G. Acoustic Echo and Noise Control: A Practical Approach. Hoboken, NJ, USA: Wiley; 2004

[23] Souden M, Wung J, Biing-Hwang FJ. A probabistic approach to acoustic echo clustering and suppression. IEEE Workshop on Applications of Signal Processing to Audio and Acoustics. 2013

[24] Hussain MS, Hasan MA, Bari MF, and Harun-Ur-Rashid ABM. A fast double-talk detection algorithm based on signal envelopes for implementation of acoustic echo cancellation in embedded systems. 4th International Conference on Advances in Electrical Engineering (ICAEE). 2017. DOI: 10.1109/ICAEE.2017.8255353

[25] Makino S, Lee TW, Sawada H. "Convolutive Blind Source Seperation for Audio Signals " in Blind Speech Seperation. USA: Springer; 2007. pp. 1-42

[26] Sawada H, Ono N, Kameoka H, Kitamura D, Saruwatari H. A review of blind source separation methods: Two converging routes to ILRMA originating from ICA and NMF. APSIPA Transactions on Signal and Information Processing. 2019;**8**:1-12

[27] International telecommunication union. G.168: 04/2015 Digital network echo canceller. Available online: https://www.itu.int/rec/T-REC-G.168-201504-I/en [Accessed: December 15, 2021]

[28] Benesty J, Morgan DR, Cho JH. A new class of doubletalk detectors based on cross-correlation. IEEE Transactions on Speech and Audio Processing. 2000; **8**(2):168-172

[29] Google Android team. 5.4.2 Capture for voice recognition. In: Android compatibility definition document. Available online: https://source.android.com/compatibility/10/android-10-cdd [Accessed: December 16, 2021]

[30] Kellermann W. Strategies for combining acoustic echo cancelation and adaptive microphone beamforming array. IEEE. 1997:219-222

[31] Hazrati O, Loizou PC. The combined effects of reverberation and noise on speech intelligibility by cochlear implant listeners. International Journal of Audiology. 2012;**51**(6):437-443

[32] Volvo Cars Sverige AB. Volvo Cars collaborates with Google on a brand new infotainment system. Available online: https://group.volvocars.com/news/connectivity/2018/volvo-cars-collaborates-with-google-on-a-brand-new-infotainment-system [Accessed: December 15, 2021]

[33] Volvo trucks Global. Volvo trucks to deliver Amazon Alexa in new heavy-duty trucks. Available online: https://www.volvotrucks.com/en-en/news-stories/press-releases/2020/dec/volvo-trucks-first-to-deliver-amazon-alexa-in-new-heavy-duty-trucks.html [Accessed: December 15, 2021]

Chapter 8

Binaural Headphone Monitoring to Enhance Musicians' Immersion in Performance

Valentin Bauer, Dimitri Soudoplatoff, Leonard Menon and Amandine Pras

Abstract

Musicians face challenges when using stereo headphones to perform with one another, due to a lack of audio intelligibility and the loss of their usual benchmarks. Also, high levels of click tracks in headphone mixes hinder performance subtleties and harm performers' aural health. This chapter discusses the approaches and outcomes of eight case studies in professional situations that aimed at comparing the experiences of orchestra conductors and instrumentalists while monitoring their performances through binaural versus stereo headphones. These studies assessed three solutions combining augmented and mixed reality technologies that include binaural with head tracking to conduct a large film-scoring orchestra and jazz symphonic with a click track; binaural without head tracking to improvise in trio or on previously recorded takes in the studio; and active binaural headphones to record diverse genres on a click track or soundtrack. Findings concur to show that better audio intelligibility and recreated natural-sounding acoustics through binaural rendering enhance performers' listening comfort, perception of a realistic auditory image, and musical expression and creativity by increasing their feeling of immersion. Findings also demonstrate that the reduction of source masking effects in binaural versus stereo headphone mixes enables performers to monitor less click track, and therefore protect their creative experience and aural health.

Keywords: headphone monitoring, binaural audio, music performance, creativity, studio recording, immersion, acoustic realism

1. Introduction

While musicians are performing on stage or in the studio, monitoring on headphones interferes with their instrument embodiment, the auditory feedback of their sound within room acoustics, and their interactions with other musicians. Indeed, wearable monitoring devices disturb the physical and technical ease that performers have acquired over a long, multi-sensory process to play their instruments or conduct ensembles at their best level. By covering their ears, headphones also jeopardize musicians' ability to control the parameters of their sound production. For instance, singers "suffer the most from the dislocation of sound that headphones engender [...] because the sound is produced in their bodies, resonating in the chest cavity and sinuses" [1]. As another example, the absence of direct auditory feedback compromises "the production of high-quality trumpet tone [that] is achieved by a combination of the correct vocal tract position, the lip-reed mechanism, and the player's breath control" [2]. Moreover, headphone monitoring obstructs collective soundscapes and established ways of listening and playing music together. To mitigate these challenges, performers sometimes remove one earcup [1] to attenuate their feeling of exclusion from the acoustic environment or to compensate for the lack of externalized sources that wearable monitoring devices as opposed to onstage speaker monitors induce [3]. In this chapter, we examine orchestra conductors' and music improvisers' experiences with wearable monitoring devices, and we discuss three binaural technology solutions that overcome stereo headphone monitoring challenges for a range of professional performance contexts.

Headphone monitoring was introduced in recording studios where it was necessary to isolate sound sources and synchronize performances on cue tracks while enabling musicians to hear themselves and others. Whereas this technology offers flexibility and creative possibilities such as overdubbing on previously recorded takes, it calls for the use of visual cues through windows and red lights, and for the setup of talk-forward and talkback microphones that may expose musicians to the others' comments on their performances. In such a technological environment for music creation, sound engineers control both the quality of headphone mixes and the communication system in the studio. Williams highlighted how the setup of the communication system increases stress and may result in tensions between musicians and engineers during recording sessions [1]. Also, adding headphone monitoring as yet another layer of engineers' sound control may worsen experiences of gendering and microaggressions in the commercial recording studio [4]. Therefore, although "the number of available headphone mixes becomes a status marker reflecting the professional standing of the studio among competing facilities" [1], using a high number of headphone mixes may negatively impact the production workflow and the social climate of the workplace. Our approach consists of adapting technologies to specific performance contexts to enhance musicians' immersion in their artistic tasks, and thus reduce stress and other adverse sociopsychological effects of headphone monitoring.

The audio content of monitoring systems influences all aspects of musicians' performances, in positive and negative ways. For instance, balancing harmonic versus rhythmic sections in a singer's or a melodic instrumentalist's monitoring mix impacts their comfort in finding their best tuning, rhythmic placement, and dynamics. Furthermore, signal processing like equalization, dynamic range compression, delays, and reverberation is commonly used to facilitate ensemble cohesion. As an example, boosting the attack of the kick drum in a bassist's monitoring mix can enhance the groove of a band. Also, a study showed that monitoring different reverberation lengths of room acoustics affects orchestra conductors' tempo, timbre, and appreciation of the performance quality when listening to recorded takes [5]. Findings from a PhD thesis about live engineering on Broadway underline how engineers are responsible for "sonic colors" that represent "the unique resonant characteristics of sound sources associated with music-making, but also to invoke "color" as a broader metaphor for social difference and identity" [6]. From this perspective, both the sound capture system and mixing approach of monitoring systems must meet the cultural expectations and genre conventions of specific performance contexts. For each of our three binaural solutions, we detail how we designed the monitoring technology, the sound capture system, and the mixing approach to satisfy the requirements of specific performance contexts.

Our interdisciplinary team of four researchers who are also experienced sound engineers and music performers aim at examining the following research questions:

1. What are the main challenges of using a wearable device for monitoring while performing music? And to what extent do these challenges differ between conducting large ensembles versus improvising?

2. Could binaural headphone monitoring technologies that are adapted to specific performance contexts enhance musicians' listening comfort, perception of a realistic auditory scene, musical expression and creativity?

3. Could binaural headphone monitoring systems decrease the click-to-music ratio compared to stereo headphones?

Before we present a fresh perspective on the methods and results of a series of three studies that were published in the proceedings of *Audio Engineering Society Conventions* [7–9], we highlight previous research on delivering synchronization auditory cues to performers; augmented and mixed reality audio applications; and binaural music production that informed our solution designs. Then, we discuss the methods and outcomes of two online surveys about orchestra conductors' and improvisers' experiences when monitoring through headphones. The survey findings serve as a basis to support the design of eight case studies that aimed to compare binaural versus stereo headphones in recording or rehearsal situations.

Because musicians rely on the auditory cues that their monitoring systems convey to elaborate their performance process, comparing the influence of binaural versus stereo monitoring on musicians' performances requires researchers to design "ecologically valid" experimental protocols and technologies that address creative cognition [10, 11]. Hence, we carried out our eight case studies in real-life performance situations.

With experienced musicians, to test three binaural monitoring solutions that we designed to meet the esthetic and cultural context of three distinct performance situations. Finally, we provide ideas for future research with audio augmented and mixed reality applications to facilitate musicians' immersion in the performance.

2. Literature review

2.1 Delivering synchronization auditory cues to music performers

The use of a click track in music performance was first documented for the soundtrack recording of *Fantasia* (Disney, 1940). Maestro Leopold Stokowski, who was an audio engineer at Bell Labs, experimented with new recording workflows to synchronize different sections of orchestra and principals on a multitrack device [12]. While the need for a click track was justified by such a creative innovation, its extensive use in music performance comes with downsides. Like sirens or fire alarms, click sounds are designed to grab attention with a lot of high-frequency energy. Therefore, long exposures to high levels of click tracks contribute to the risks of musicians' hearing loss [13]. Although click samples can be changed in digital audio workstations to accommodate musicians' preferences, the mechanical nature of the click implies that "overall, playing with a click track means playing with the metronome" [14]. According to Cardassi [15], "a click track is likely the most dreaded synchronization tool in music," and it generates performers' "angst and unpleasantness." Drawing upon Blauert [16]'s theory, spatial audio applications offer greater source discrimination possibilities than a stereo image. Therefore, binaural technologies provide sound engineers with more mixing space than stereo, which implies more source-positioning options and the need for less equalization and dynamic range compression [17] to avoid masking effects among sound sources. Consequently, we suggest that binaural headphone monitoring solutions allow for lower click track levels and less processed instrumental and vocal sources for performers to synchronize with each other, on a soundtrack or a movie, meanwhile protecting their aural health and improving their creative experience.

Previous research suggests that a generalized use of click tracks has homogenized creativity and globalized music cultures. For instance, an analysis of tempo across the past 60 years of U.S. Billboard Hot 100 #1 Songs revealed that a 5-beat average standard deviation from 1955 to 1959 decreased to 1-beat between 2010 and 2014 [18]. Moreover, Éliézer Oubda, a music producer and sound engineer who owns *Hope Muziks Studio* in Ouagadougou, Burkina Faso, trains his assistants in explaining to Western African musicians how to perceive the downbeat in the click track in the same way Europeans and North Americans do.[1,2] To minimize "the straightjacket feeling" [14] induced by click tracks, composers, performers, and studio professionals can collaborate on developing alternative cue tracks and monitoring systems. For instance, customized tracks may combine pre-recorded fragments from the parts to be performed with vocal instructions or relevant pitches. These may also feature excerpts of embedded click tracks within the pre-existing layers of audio to provide additional guidance at specific times only. These cue improvements can be accompanied by a context-dependent choice of the monitoring system. While high-fidelity technologies may not always be the best solution[3], the selection of a wearable device requires some attention. Typical studio headphones consist of closed-back headphones that are "designed to block out environmental noise using a passive acoustic seal" [19]. Mostly found in live scenarios, in-ear monitors provide a more drastic acoustic isolation, with visual discretion and stability benefits in situations where the performer frequently moves their head. With non-isolated ear cups, open-back headphones offer a more natural or "speaker-like sound" [19] with a more pleasant spatial image, and less risk of performers feeling isolated and disconnected from the environment. Whereas we did not consider using open-back headphones for our monitoring applications because their audio content would leak into the microphones, their benefits have inspired our binaural solutions to overcome the auditory feedback challenges of stereo closed-back headphones and in-ear monitors.

Two types of technologies exist to deliver synchronization auditory cues to musicians while providing them with direct access to their own sound production and acoustic environment, namely acoustic-hear-through and microphone-hear-through monitoring systems [20]. Primarily developed to improve the safety of outdoor runners when they are listening to music, acoustic-hear-through monitoring systems, also known as bone conduction headphones leave the users' ear canal free by conveying the auditory cues "from the vibration of the bones of the skull [or jaw] that is transmitted to the inner ear" [21]. Whereas this technology eliminates

---

[1] See 15:00-20:10 of the roundtable discussion about "De-colonizing the Digital Audio Workstation" with Éliézer Oudba and Eliot Bates facilitated by Menon and organized by Pras and Kirk McNally: https://www.canal-u.tv/chaines/afrinum/roundtable-discussion-about-de-colonizing-the-digital-audio-workstation

[2] During a jembe workshop taught by Issa Traoré alias Ken Lagaré, an arranger and sound engineer who owned *Authentik Studio* in Bamako, Mali, graduate student Leo Brooks and percussion instructor Adam Mason explained the fact that European and North American musicians struggle to hear the downbeat in Western African music (see 45:50–47:10): https://www.canal-u.tv/chaines/afrinum/percussion-workshop-with-ken-lagare

[3] For example, Oubda gave the example of rural musicians from Burkina Faso who got intimidated when they heard themselves through high-fidelity headphones for the first time (see 17:00–17:40): https://www.canal-u.tv/chaines/afrinum/roundtable-discussion-about-de-colonizing-the-digital-audio-workstation

disconnection feelings from the acoustic surroundings, like open-back headphones, the monitoring mix may leak into the microphones. Indeed, May and Walker [22] reported "approximately 12 dB A (total) of 'leakage'" in the context of listening tests. Also, Cardassi, who tested a bone conduction headphone to record an electro-acoustic album on piano and vocals, could only use it for pieces that did not require the use of a close vocal microphone, and whose cue tracks did not include any click.[4] Primarily used as hearing aids devices, microphone-hear-through monitoring systems consist of mounted microphones on the users' headset that capture what they would hear without headphones [23]. Cooper and Martin [2] designed a microphone-hear-through monitoring system named Acoustically Transparent System (ATH) that combines the binaural rendering of the signal captured from two headset-mounted microphones with the synchronization cues. In performance situations, they observed that the ATH has "a notable impact on both quality of tone production and the confidence of the [trumpetist]" [2]. Their findings confirm the relevance of designing binaural technologies to improve musicians' experience while performing with headphone monitoring.

2.2 Mixed and augmented reality applications with binaural technology

Audio Mixed Reality (AMR) applications aim at recreating new auditory spaces for listeners by balancing the proportion of real and virtual elements. Also, Audio Augmented Reality (AAR) applications aim at achieving listeners' experiences of acoustic transparency, as if there was no headset, to interleave virtual sounds with an unaltered reality [24]. Drawing upon Milgram and Kishino [25]'s "virtuality continuum" of visual displays, McGill et al. [26] define AAR as "auditory headset experiences intended to [...] exploit spatial congruence with real-world elements." From this perspective, AAR sits at the edge of AMR that encapsulates "any auditory VR and AR experiences." These definitions mirror the recording esthetics continuum from "attempting realism" to "creating virtual worlds" produced through different sound capture systems and mixing approaches [27]. While mixing for stereo recordings differs from mixing for AMR and AAR applications, we applied our knowledge of sound capture systems to best meet the cultural expectations and genre conventions of the performance contexts. Specifically, we primarily used microphone arrays that captured the acoustic environment for our five AAR case studies, versus close mono microphones that focused on the instruments' direct sound for our three AMR case studies.

To enhance listeners' perception of auditory spaciousness through headphone monitoring, König [28] conceptualized one of the first four-channel headphones that positioned an additional speaker driver near the tragus to diffuse reverberation, and thus allow for a more accurate spatial image with less sound pressure level on the ear axis. Further developments intending to simulate surround and multi-channel loudspeaker systems have led to the design of multi-driver headphones that position multiple speaker drivers within the ear cup, employing the shape of the listener's ear and pinna to influence the filtering of high frequencies as they enter in the ear canal [19]. Meanwhile, most of today's AAR and AMR headphone applications use binaural filtering with head-related transfer functions (HRTF) that enable listeners to externalize sound sources while wearing regular headphones. Theoretically, delivering accurate intelligibility, localization, and externalization

---

[4] Cardassi first tested a bone-conduction headphone in Fall 2017 for the recording of *Ramos* (Redshit, 2019) with Pras as music producer and sound engineer. While she could only use it for the recording of a few pieces in Rolston Hall at the Banff Centre, she enjoyed preparing for the sessions with it at home. This was confirmed through personal email communication on March 9, 2021.

of sound sources through headphones requires the binaural rendering of sound sources via individualized HRTFs transmitted through high fidelity open-back headphones [29]. Nevertheless, according to a review of sound externalization studies, adding reverberation-related cues, and/or dynamic binaural rendering that matches listeners' self-initiated head movements, facilitates the localization and externalization of binaural cues [30], which may compensate for the use of non-individualized HRTFs and closed-back headphones. Whereas dynamic binaural ensures the success of AAR applications for users who move a lot in the real-world environment, such as orchestra conductors, we suggest that static binaural may be more relevant for AMR applications where most of the binaural cues are out of sight, such as recording sessions with musicians performing in separate rooms. In this view, static binaural might still provide users with a better source intelligibility and a more spatial experience compared to stereo systems since there is less masking effect among sources, even though the localization accuracy and externalization of binaural cues remain compromised, for example, generating front-back confusions. In fact, a study showed that "short training periods involving active learning and feedback" facilitate listeners' ability to externalize sources while using binaural systems with non-individualized HRTFs [31]. In this chapter, we present the concept of two distinct dynamic binaural AAR setups and one static binaural AMR setup that involved a short training tutorial for listeners.

Besides the popularity of noise cancelation headphones that filter the real acoustic environment out for listeners to focus on music or other virtual elements [26], AAR and AMR microphone-hear-through devices are primarily developed for single users' experiences in non-musical applications, for example, for audio gaming [32]; street navigation [33]; and soundwalks that immerse listeners in sonic art compositions [34]. Only a few collaborative AAR experiences have been tested [35], for example, a four-player interactive audio experience [36]; a two-player audio game called *eidola multiplayer* [37]; and creative artworks dedicated to multi-users, such as *Listen* for museum visits [38] or *SoundDelta* devoted to large public outdoor events [39]. Also, to our knowledge, very few AAR musical applications besides Copper and Martin's ATH [2] have been designed. For instance, a Master thesis showed that members of a rock band preferred performing with AAR dynamic setups compared to mono and stereo headphones [40]; a study with methodological shortcomings tested AAR dynamic in-ear monitors for members of an acoustic ensemble [41]; and the *Architexture Series* brought new music composers, sound engineers, and architects to collaborate on site reconstruction [42, 43]. Our eight music performance case studies, therefore, contribute to AAR and AMR research by assessing two AAR setups that aim at overcoming performers' social interaction challenges when wearing headphones, and one AMR setup that aims at enhancing social interactions among performers when being remotely located.

2.3 Binaural music production

Sound engineers increasingly use binaural technology in the recording studio in parallel with the development of new plugins and devices that enable listeners' sound externalization on headphones with and without the tracking of their head movements, for example, binaural simulation of surround sound mixes in control rooms that do not have a 5.1 speaker system [44]. Although binaural audio is optimized for headphone listening which is the primary music listening mode of our time, so far only few binaural music productions have been released on the market. For instance, Williams and Reiser walked us through the binaural capture and rendering processes of sources for the production of "*GoGo Penguin [untitled]*"

(Blue Note Records 2020), which was released in stereo and not yet in binaural.[5] They used three *Neuman KM 100* dummy heads to *overdrive* space in the main live room and the drum room, and to immerse listeners within the piano sound. At the mixing stage, they also used *dear VR* plugins to externalize specific sources. They underlined that binaural production techniques are the best fit to convey virtuosic performances of high-level musicians in contemporary jazz and classical music because the recording of their performances requires little signal processing in terms of equalization and dynamic range compression. Indeed, extensive signal processing does not work well with binaural rendering, and equalization and compression should only be used for creative purposes since there are less source masking effects than in stereo [17]. We thus assessed our three binaural solutions in professional-level performance contexts whose esthetics did not require much signal processing, with five out of the eight case studies primarily involving classical and jazz musicians.

Whereas binaural has not yet succeeded commercially as a release format, more and more public European radios offer binaural programs, for example, *Hyperradio* on Radio France, which primarily broadcasts audio plays and electronic music live shows. To broadcast classical orchestral recordings for *BBC Proms* on BBC Radio 3, Parnell and Pike [45] reported on using IRCAM's *Panoramix* to enhance the positioning and ambiance of the auditory scene captured with a Schoeps ORTF-3D microphone array that features two coincident layers of four microphones. Results from their audience study showed that binaural mixes were rated as "more enjoyable" by 79% of respondents, whilst 75% said that the experience was "somewhat" or "absolutely" like being there in person. These findings contrasted with previous research that found that overall, the stereo listening experience was preferred to binaural for a range of musical genres [46]. Also, the outcomes of a study about binaural mixing for hip-hop production suggest that listeners can be disoriented by this unfamiliar immersive format [47]. In particular, the main sources of the beat seem more effective when not externalized. We used this knowledge to capture and mix sound sources in the performers' binaural headphones for our eight case studies.

3. Online survey on music performers' experiences with headphone monitoring

3.1 Online survey methods

A combination of two online surveys further examined the challenges that music performers face when wearing monitoring devices in the studio or on stage [7, 8].

3.1.1 Respondent demographics

We recruited 12 orchestra conductors and 12 music improvisers from our respective networks by email to fill out a survey on an unpaid, volunteer basis. These 24 professional respondents included 20 males and four females living in seven countries (Australia, Canada, France, Germany, Netherlands, Switzerland, the UK, and the USA). They had at least 5 years on the job, except for one who reported having between one and 5 years on the job. More than half (15 out of 24)

---

[5] https://mupact.com/seminar-program-may-jul-2020/aesthetic-manifestos-and-binaural-integration-an-investigation-of-pre-in-session-and-post-production-techniques-employed-in-gogo-penguins-self-titled-2020-album-release/

were touring internationally; the other nine were primarily working in France. All 24 respondents had experienced headphone monitoring while performing. Half of the conductors primarily performed for studio recording sessions with acoustically isolated instruments and/or the need to overdub on previous recordings; five for live concerts of film-scoring or new music compositions with electronic components; and one for both kinds of performance situations. Nine of the improviser respondents reported wearing headphones for more than half of their studio recording sessions; and three of them for 30% or less. Improvisers played a variety of instruments and included a singer; an acoustic bassist; a trombonist; a hornist; two saxophonists; one flutist and electronic artist; one multi-instrumentalist who played sousaphone, saxophone, clarinet, and flute; two drummers (one also conducted ensembles); and two pianists (one also played electronic keyboards and produced recordings, the other one also sang and played prepared piano). About musical genres, improvisers primarily performed jazz and/or world music (53%); pop-rock subgenres including French *variété* (27%); experimental, improvised, or contemporary music (9%).

3.1.2 Questionnaire

Both surveys used similar semi-directed questionnaires because Bauer et al. [8] adapted Soudoplatoff and Pras [7]'s methods from the context of orchestra conducting to the context of music improvisation. In this chapter, we focus on the analysis of the respondents' answers to four questions that were featured in both questionnaires. These questions are slightly reworded here to encompass both performance contexts (i.e., orchestra conducting and recording improvisations):

1. According to your previous studio recording session experiences, how would you describe an ideal headphone monitoring system?

2. Think about one of your best studio recording sessions with headphone monitoring. Start by describing the context of this session (ensemble, production, location, etc.). Why do you think this session was a success?

3. Think about one of your worst studio recording sessions with headphone monitoring. Start by describing the context of this session (ensemble, production, location, etc.). Why do you think this session went this way?

4. When recording in the studio, do you have a particular way of wearing headphones? If so, why?

3.1.3 Qualitative data analysis

Respondents' verbal descriptions were analyzed using a *Grounded Theory* approach [48] drawing from previous research on studio practices (e.g., [49]). This approach consists of extracting meaningful phrasings from the free-format verbal descriptions to be classified into concepts and categories without preconceived themes. Specifically, the constant comparison technique of Grounded Theory called for a minimum of two researchers to review each other's classification and to draw parallels between findings from the different questions, to gradually refine the emerging concepts and categories as well as to identify consensus and contradictions among outcomes of different questions. Results will be presented with the count of phrasing occurrences for each concept and category.

3.2 Online survey analysis

3.2.1 Ideal headphone monitoring system

We identified 50 phrasings from the respondents' free-format verbal descriptions of their ideal headphone monitoring system. These phrasings were classified into three major categories, namely Sound Quality (n = 24), System Technical Quality (10), and Physical Properties (10); and into three minor categories, namely Click (3), Ambiance in the Studio (2), and Forgetting the Headphones (1). The most-reported concepts for each major category were Realism (8), Instrument balance (8), and Control over monitoring (6). **Figure 1** displays the classification into emerging concepts and categories of the 31 phrasings coming from improvisers and the 19 phrasings coming from conductors separately, since a Yates' chi-squared test revealed a significant difference between the answers' distribution into the six categories for conductors and improvisers (χ^2 (5,50) = 2,97, p < 0.05).

3.2.2 Positive and negative experiences when performing with headphones

In total, we collected 129 phrasings, 70 from improvisers and 59 from conductors about their positive and negative experiences when performing with headphones. A Yates's chi-squared test revealed no significant difference between the answers' distribution into the nine categories for conductors and improvisers (χ^2 (8,129) = 15,91, p > 0.05). Nevertheless, we chose to keep both populations of performers distinct in **Figure 2**, to stay consistent with the

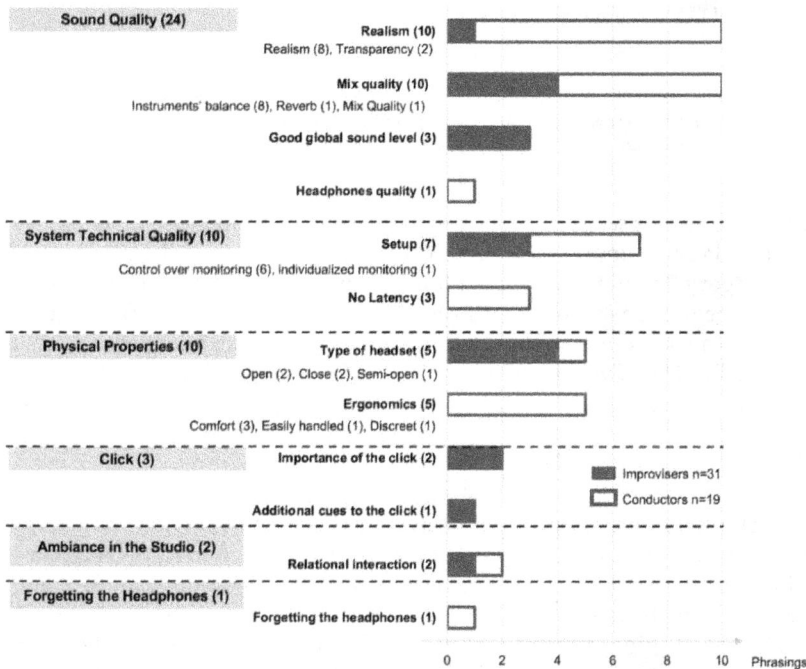

Figure 1.
Classification of phrasings extracted from the 12 improvisers' and 12 conductors' free-format verbal descriptions accounting for their ideal headphone monitoring system.

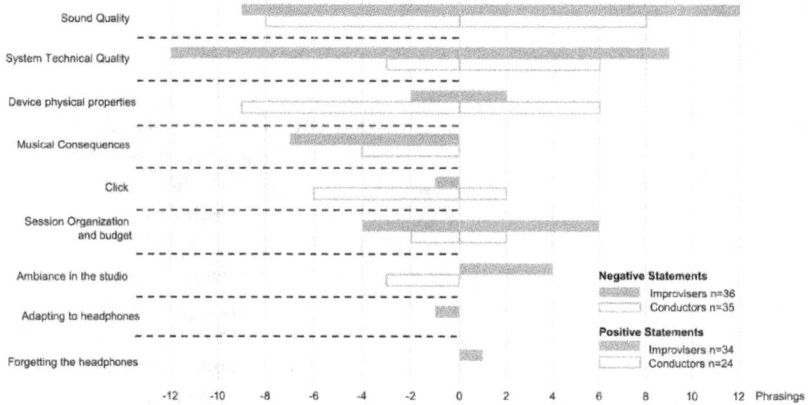

Figure 2.
Classification of phrasings extracted from the 12 improvisers' and 12 conductors' free-format verbal descriptions accounting for their positive (green) versus negative (orange) monitoring experiences.

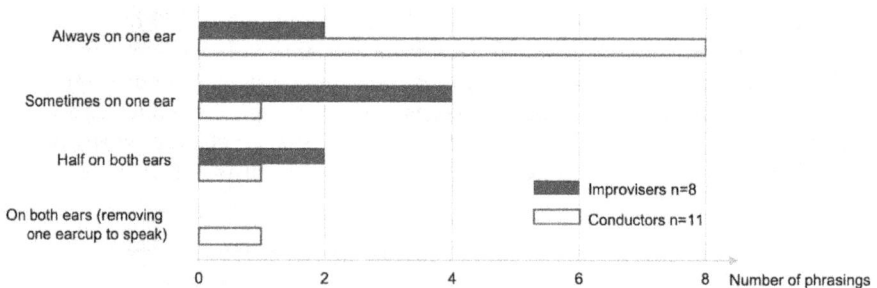

Figure 3.
Classification of phrasings extracted from the 12 improvisers' and 12 conductors' free-format verbal descriptions accounting for their usual ways of wearing headphone monitoring devices.

other figures in this section. Regarding positive experiences, 58 phrasings were identified, 34 from improvisers and 24 from conductors, and classified into the major category Sound Quality (20), followed by System Technical Quality (15). Regarding negative experiences, 71 phrasings were identified, 36 from improvisers and 35 from conductors, and classified into three major categories, namely Sound Quality (17), System technical Quality (15), and (negative) Musical consequences (11).

3.2.3 Ways of wearing headphones

We collected 19 phrasings, eight from improvisers and 11 from conductors about their ways of wearing headphones. A Yates's chi-squared test revealed a significant difference between the answers' distribution into the four different habits of wearing headphones for improvisers and conductors (χ^2 (3,19) = 6,42, $p < 0.05$). Hence, **Figure 3** presents the classification of phrasings for the improvisers and conductors separately. The main habit that we identified consisted in always (Improvisers: 2; Conductors: 8) or sometimes (I: 4; C: 1) wearing the device on one ear only.

4. Assessment of three binaural headphone monitoring technologies in a performance situation

4.1 Technology design and experimental protocols

We designed three binaural headphone monitoring solutions to enhance musicians' cognitive engagement in performance (**Table 1**). For each solution, we adapted the augmentation type, sound capture system, and mixing approach to the esthetic and cultural context of distinct performance situations (**Table 2**). Then, we conducted eight case studies that involved two renowned conductors of symphonic ensembles with a click track in large acoustics [7]; seven emerging music improvisers in solo or trios in separated dry rooms or overdubbing alone in small acoustics [8]; and three music students and one touring musician who recorded for a range of musical genres alone or in a duo with a click track or a soundtrack in medium-size acoustics [9]. These eight case studies were all carried out in real-life performance situations at the Paris Conservatoire (CNSMDP), Radio France, and the University of Lethbridge (ULeth).

Our mixed methods of assessing these solutions draw upon Agrawal et al. [50]'s definition of *immersion* as a psychological state that enables an individual's mental absorption in the world and in the tasks that are presented to them. Therefore, for each performance case study, we determined which auditory information would be the most important for the users to monitor in order to perform at their best, in other words, which auditory information would be "immersive enough" [51] to achieve a sense of "being there together" [52].

4.1.1 Description of three binaural headphone monitoring solutions

Table 1 highlights the binaural rendering pipelines and augmentation technologies that we chose to best adapt to performers' needs for each context. To enable conductors to monitor large ensembles on headphones, Soudoplatoff and Pras [7] designed a Binaural with Head Tracking (BHT) system that rendered a *JML* tree [53], that is, a main five-microphone array with specific dimensions, and integrated spot microphones. This system used *Bipan*[6] software [54] coupled with *Hedrot*,[7] that is, a head tracker located on the conductors' headphones. In Bipan, the LISTEN database [55] was used with the HRTF pair n°1040, as advised in a previous study [56], since this HRTF pair satisfied most users during public demonstrations of the software [57]. Bipan had a latency of 5.3 ms when used with a buffer size of 256 samples. According to previous research, a monitoring system latency below 42 ms should be acceptable [58]. Furthermore, Hedrot had a latency of 48.1 ± 4.3 ms [54], which should provide conductors with accurate localization cues since the head tracking latency does not hinder the stability of virtual sounds within complex auditory scenes under 71 ms [59], even if it could be noticeable when superior to 30 ms [60]. The assessment tests required the use of the *TotalMix* application, which has a meaningless latency of three samples (equal to about 68 μs at 44.1 kHz), to digitally convert the microphone signal and send it to the BHT via a *RME MADIface*.[8]

---

[6] 3D audio technology developed in-house at the Paris Conservatoire in collaboration with IRCAM as part of the Bili project: http://www.bili-project.org/. More details can be found here: https://alexisbaskind.net/fr/bipan-binaural/

[7] https://abaskind.github.io/hedrot/

[8] https://www.manualslib.com/manual/1310692/Rme-Audio-Madiface-Usb.html?page=70

Performance context		Binaural rendering pipeline		Augmentation related to the context		
Ensembles [# users]	Sync. cues	Microphone system	Binaural Rendering	Acoustic venus	Type of immersion	Static vs. dynamic
BHT 2 Conductors of large Ensembles [2]	Click & Rhythmic section	5-microphone array + spot mics	Bipan with LISTEN HRTF pair 1040 + Hedrot head tracker with latency of 48.1 ± 5.3ms	Large acoustics	AAR	Dynamic
BMR 1 Solo & 2 Trios [7]	Overdubbing & Rhythmic section	Close mics	KF with proprietary anechoic HRTFs	Isolated dry rooms	AMR	Static
ABH 2 Soli & 1 Duo [4]	Click or Soundtrack	2x2-mounted microphones	KV with proprietary anechoic HRTFs	Medium acoustics	AAR	Dynamic

Table 1.
Performance context, binaural rendering pipeline, and augmentation principles of the three binaural headphone monitoring technologies.

To enhance the intelligibility of improvisers' subtle expressive gestures, Bauer et al. [8] developed a Binaural Mixed Reality (BMR) system that rendered close mono microphones through KLANG: fabrik (KF) hardware. KF was chosen for its convincing externalization of sources and sound quality[9] as well as its latency of less than 3 ms.[10] Indeed, the set of KLANG-proprietary HRTFs was preferred to HRTFs from the LISTEN database that features a low sampling resolution, introduces noise artifacts, and present amplitude errors, for example, for the HRTF pair of subject IRC_1034 [61]. The BMR had a total latency (KF latency plus ProTools latency) of 4 ms for the two trios. Regarding the world music performer, the technical setup between the microphone signal and the monitoring system included several digital devices, and the measured total latency of the chain was 14.1 ms. The musician specified that he did not notice it, and confirmed that the system latency did not hinder his performance.

To attempt acoustic transparency of the recording auditory space, Menon [9] built an Active Binaural Headphones (ABH) system with two 150°-angled small condenser microphones mounted on each earcup. Based on Bauer et al. [8]'s satisfying findings, the signal coming from the four mounted microphones was binaurally rendered through KLANG: vier (KV) hardware, which features the same sonic and latency properties than KF[11]. The ABH total latency was inferior to 16.8 ms. The assessment tests required the use of the *CueMix* application that has no latency to digitally convert the microphone signal and send it to the ABH via a *MOTU 896*

---

[9] The researchers were able to evaluate the quality of this equipment as they are experienced sound engineers, in both stereo and 3D audio production techniques.

[10] https://www.klang.com/en/products/klang_fabrik

[11] https://www.klang.com/en/products/klang_vier

$mk3^{12}$ that has a latency of under 13 ms, and the *Aviom* personal monitor mixer that has a latency of 0.88 ms to amplify the headphone signal.[13]

In summary, the BHT and ABH are two AAR systems with dynamic binaural because for both of their applied contexts, performers primarily needed to monitor sound sources while being in the same room as their peers, and thus required a technology that accurately conveyed source localization. On the other hand, the BMR is an AMR system with static binaural because improvisers primarily needed to monitor their previous recordings or their band members who were playing in separate rooms; thus, the re-creation of a virtual space that facilitated their immersion was more desirable than accurate source localization.

For all three technologies, closed-back headphones were used to minimize sound leakage into the microphones. Both Bipan and KLANG used anechoic HRTFs, and so enabled us as sound engineers to generate spatial images with re-created acoustics that fit the acoustics of the performance space.[14] These HRTFs were also non-individualized and thus required performers' listening training [31] and/or dynamic binaural rendering [29] to optimize source externalization and mitigate timbre artifacts. Therefore, one week before conducting the case studies that assessed the BMR setup, which is static, the improviser participants were instructed to listen to three-to-five binaural audio productions over headphones that were selected from *Hyperradio* podcasts by Bauer (total duration of around 25 mn), to get used to the binaural rendering. All of them confirmed to Bauer at the beginning of their recording session that they had listened to at least three of these productions. This consists of a total listening experience of 15 mn at minimum for each participant.

4.1.2 Case study procedures for binaural solution assessment

Table 2 details the locations, genres, and instrument line-ups of the eight case studies in chronological order for testing our BHT, BMR, and ABH technologies in rehearsal or studio recording situations. Thirteen performers agreed to participate in these comparative tests without financial compensation. The first two tests that involved symphonic ensembles were organized at the institutional level as part of a pedagogical project. For the other five tests, Bauer and Menon volunteered to mix the recordings, which the performers could use to promote their music.

To assess the three headphone technologies that are described in the previous section, two conductors, seven improvisers, and four musicians who perform a range of musical genres compared binaural against *traditional* stereo headphones, that is, the monitoring systems commonly used in each of the performance venues. The experimental procedures for each case study are summarized in **Table 2**. Because "an experimental protocol is ecologically valid if the participants react [...] as if they were in a natural situation" [10], Soudoplatoff organized the first two case studies during rehearsals of programmed productions with large ensembles. Specifically, for the last two days of a week of film-scoring rehearsals, Maestro Laurent Petitgirard agreed to swap headphone conditions five times during breaks that occurred every 90 min, which led him to test each condition three times. Unfortunately, the comparison could not be carried out with the jazz symphonic ensemble due to a conjunction of acoustic and organization issues (see Section 4.2

---

[12] https://motu.com/techsupport/technotes/what-is-the-latency-of-my-motu-audio-interface

[13] https://www.aviom.com/library/User-Guides/36_A-16D-User-Guide.pdf

[14] Using non-anechoic HRTFs implies generating a binaural image that emulates the externalization of sources in specific room acoustics. This may be enjoyable for the listener and can be creative in the context of music production. However, it is likely to be confusing for the musician in the context of headphone monitoring when performing.

	Institution	Venue	Audio latency	Musical genres	Instruments	Performance purpose	Comparison procedure [# cases; duration]
BHT	CNSMDP	Art lyrique	5.3 ms	Film-scoring	Symphonic orchestra	Rehearsal	B S B S B S [6; 90 mn]
		GPO	5.3 ms	Symphonic jazz	Symphonic orchestra with a jazz big band and non-acoustically amplified instruments	Studio recording	*Comparison not possible*
BMR	Radio France	Studio 115	14.1 ms	World music	Voice, bass, various percussions, small guitar	Studio overdubbing	B S [2; 2 h]
	CNSMDP	240/244/245	4 ms	Jazz trio	Double bass, drums, electric guitar	Studio recording	B S [2; 45 mn]
			4 ms	Free improvisation trio	Drums, clarinet/bass clarinet, accordion	Studio recording	S B[2; 45 mn]
ABH	ULeth	Studio one	3.8 ms	Singer-songwriter	Voice & banjo	Studio recording	S B [2; When musicians were pleased with the stereo takes]
			3.8 ms	Pop-rock	Drums & electric guitar	Studio recording	S B [2; When musicians were pleased with the stereo takes]
		Recital hall	3.8 ms	Electroacoustic	Piano and acoustically amplified soundtrack	Studio recording	S B [2; When musicians were pleased with the stereo takes]

Table 2.
Location, genre, instrumentation, and comparison procedure of the eight case studies—B refers to the binaural condition and S to the stereo condition.

for explanations). Bauer and Menon ensured the ecological validity of their experiments by inviting performers to record in the studio with the incentive of getting a demo that they could use to promote their music. In this context, the world music performer and two improvisation/jazz trios accepted to test the BMR system in a counterbalanced order, and each switched conditions once, after 2 h and 45 min, respectively. Also, a singer-songwriter, a rock duo, and a pianist who performed with electronics accepted to test the ABH system once they were satisfied with their takes using the traditional stereo system of the studio.

For the seven case studies during which performers compared binaural and stereo headphones, the researchers took notes on users' behaviors and comments during the tests. Whereas Soudoplatoff asked Maestro Petitgirard to react spontaneously after each trial, Bauer conducted post-test focus group interviews, and Menon carried out individual post-test written surveys with the performers at the end of the recording session. For all case studies, performers were asked to compare both types of headphones in terms of comfort, playfulness, benchmarks, and perception of the spatial image. For the recording sessions only, performers were asked to compare the perception of their own instrument in relation to others'. Moreover, a few weeks after the recording sessions of the world music performer and the two improvisation/jazz trios, Bauer sent stereo mixes of all the takes to the performers, and he asked them to select their favorite take for each piece (or their favorite improvisation). Based on a previous performance study in the recording studio in jazz [62], collecting musicians' choice of takes that were recorded in different conditions has the potential to inform the impact of the BMR on creativity and musical results. The context of Soudoplatoff's and Menon's tests did not allow for this additional collection of data.

4.1.3 Click-to-music loudness ratio measurements

To investigate the extent to which the reduction of the sound masking effect in binaural enabled musicians to monitor less synchronization cues, for each of his three case studies[15], Menon [9] compared the click-to-music (CMR) loudness ratio between the headphone mix recordings of the takes using his ABH and those using the traditional stereo monitoring system of the studio. For each of the takes recorded with the ABH, he copied the musicians' KV interface settings into a "second user," so that he could print the monitoring mix that featured the binauralization of the four headphone-mounted microphones and the synchronization cues. For each of the takes recorded with the stereo headphones, he captured the signal from the headphone output of the Aviom personal monitor mixer by using a stereo jack into two unbalanced jack adapters and two Direct Input (DI) boxes. Then, because each monitoring mix replica would include a few seconds of synchronization cues before the beginning of the music performance, he could normalize the loudness of each replica with the synchronization cues as a reference. This data acquisition procedure enabled the visualization of the CMR throughout and across takes.

4.2 Experimental findings

For the seven case studies during which performers compared binaural and stereo headphones, all performers favored the binaural over the stereo condition. In the following sections, we detail comparison findings for the main criteria that emerged from our analysis of performers' comments and take choices, namely Listening comfort; Perceived realism; and Musical expression and creativity.

For the symphonic jazz ensemble recording session, the comparison could not be conducted as planned due to several challenges that highlighted the limitation of the BMH [7]. This large ensemble combined orchestral and big band instruments with electric guitars and keyboards that were not amplified in the room, as well as drums that were semi-isolated in the room. Consequently, the electric guitars, keyboards, and the double-bass' quiet acoustic sound were not captured by the main 5-microphone array so they could not be homogeneously integrated into the auditory scene. Also, the main array captured a lot of drum leakage, which damages the intelligibility of the auditory scene. Moreover, the

---

[15] Menon [9] conducted a fourth case study with a classical pianist who tested the ABH and compared it to stereo headphones to monitor a metronome while performing Beethoven's Piano Sonata No. 2, Op. 57. Because this piece would not be performed with a metronome in professional situations, we excluded this fourth study from this chapter.

complexity of the situation generated communication challenges between the electric instrument players, the sound engineer, and the conductor, therefore the conductor did not feel comfortable enough to use the BHT for the session. In the discussion, we provide ideas to overcome the BHT limitations for conducting large ensembles that blend different types of instruments in large acoustics.

4.2.1 Listening comfort

All eleven performers who participated in comparative studies in the recording studio preferred the auditory feedback quality of their own sound production in the binaural conditions. In particular, two improvisers who tested the BMR and all performers who tested the ABH reported having more control of their own instruments. For instance, the world music performer kept both earcups in the binaural condition but removed one earcup to control his voice in relation to the room acoustics in the stereo condition. Also, the double bass player of the jazz trio perceived a more realistic "physical-auditory contact" with his instrument in the binaural condition.

The conductor and seven instrumentalists expressed being more comfortable while performing in the binaural condition. In particular, whereas Maestro Petitgirard was a bit reluctant to try the BHT in the beginning, he mentioned feeling comfortable with it as soon as he started using it. Also, two out of the four performers who tested the ABH stated that they were able to forget about the device while monitoring in binaural. Furthermore, three out of the seven performers who tested the BMR reported that the binaural condition was less tiring in comparison with the stereo condition. Only the world music performer was disturbed during the first hour by this new kind of monitoring.

All performers perceived better sound quality in the binaural condition that they described as more natural than stereo in terms of spatial realism and audio clarity. With the BMR, all performers perceived the binaural mix as more intelligible, since they could better differentiate the details of the different instruments. In this view, free improvisers and jazz musicians reported "not having to force" to hear what they needed to react to their bandmates' musical gestures. They could appreciate more subtleties in their playing, for example, the sounds of the fingers on the double bass and soft percussions, and the drummer said that the sound was more "accurate to what they would hear in their daily practice." Also, the free improvisers who used the BMR and Maestro Petitgirard who used the BHT perceived more depth in the binaural mix compared to the stereo mix.

4.2.2 Perceived realism

Across the seven comparison studies, performers expressed that binaural monitoring was more realistic. However, the meaning of *realism* varied according to the type of augmentation that was used in the different studies. Regarding the two AAR systems, realism implied that the binaural rendering of the music signal was close to the real auditory environment in terms of source spatialization, room acoustics, and timbre quality. In contrast, regarding the BMR solution that is AMR, by realism performers meant that they could recreate familiar auditory situations in their mind, for example, to "be in the performance" and to connect with other players and their own instrument like in rehearsal. In the next paragraph, we illustrate these two meanings of realism with test observations.

When first trying the BHT, Maestro Petitgirard thought that he was only hearing the click track, and Soudoplatoff had to convince him that the orchestra was also rendered in the headphone mix by muting the microphones for a few seconds. Similarly, all performers who experimented with the ABH mentioned that they perceived a more realistic spatial image in comparison with stereo monitors. Beyond the basic acknowledgment of the spatial authenticity that the ABH facilitated, performers commented explicitly on the efficacy of this enhanced acoustic realism. For instance, the pianist who performed the electroacoustic piece stated, "I felt myself making decisions in real-time, reacting to my own emotions and improvising some aspects of interpretation, whereas with the traditional headphones, I found my performance becoming stagnant." As for the AMR system, since 3D audio cues did not match the real auditory scene of the studio, realism was about the sound quality of recreated acoustics and the convincing spatialization of 3D audio cues. This led the world music performer to report that he "had the impression that the music was real around him." Moreover, two of the free improvisers had the impression that their bandmembers were next to them although they were in separated rooms. In particular, the clarinetist said: "It recreated a second room where we were all present in my head."

4.2.3 Musical expressivity and creativity

All performers who tested the BMR or the ABH stated that binaural monitoring positively impacted their musical playfulness and creative process. Whereas performers did not expand verbally on this impact, six out of the seven who used the BMR only selected takes that were recorded in the binaural conditions. Also, we observed that the takes that were recorded by the free improvisation trio with binaural monitoring lasted longer, and the clarinetist reported, "musical ideas came faster." Moreover, the guitarist of the jazz trio expressed being able to take more risks, and the world music performer reported being inspired by the binaural auditory space to build his composition in the studio. In contrast, whereas Maestro Petitgirard perceived the BHT as very pleasing, he said that the monitoring condition should not have impacted his way of performing as he had drawn well-established habits over years of conducting experience.

The free improvisers who used the BMR and all performers who used the ABH expressed that they performed more intimately in binaural conditions. For instance, the free improvisers noticed that they performed the only soft improvisation with many subtleties while monitoring in binaural. Similarly, the pianist who played an electroacoustic piece said that binaural monitoring facilitated a more sensitive performance. Moreover, synchronization cues were more easily perceived in the binaural condition. Indeed, the singer-songwriter who tested the ABH noted that keeping tempo was easier, and the drummer who used the BMR reported that there was better bass/drums cohesion in the binaural condition, which led to more swing.

4.2.4 Click-to-music loudness ratios

The Click-to-music-ratio (CMR)[16] analyses were measured in relative Loudness Units (LU). These analyses across tests showed that the CMR was 4.2 LU to 17.4 LU lower when using the ABH compared to the stereo systems [9]. **Figure 4** displays the CMR in LU at key performance moments of the pop-rock duo for the drummer's monitoring mix with a click track (A), and at key performance timings of the electroacoustic piece for the pianist's monitoring mix with a soundtrack that included

---

[16] Examples of the binaural versus stereo monitoring mixes that the musicians heard are available under this link: https://www.youtube.com/watch?v=8c8lBCzJR-M

(A)

—●— Stereo Headphones   ⋯●⋯ ABH

(B)

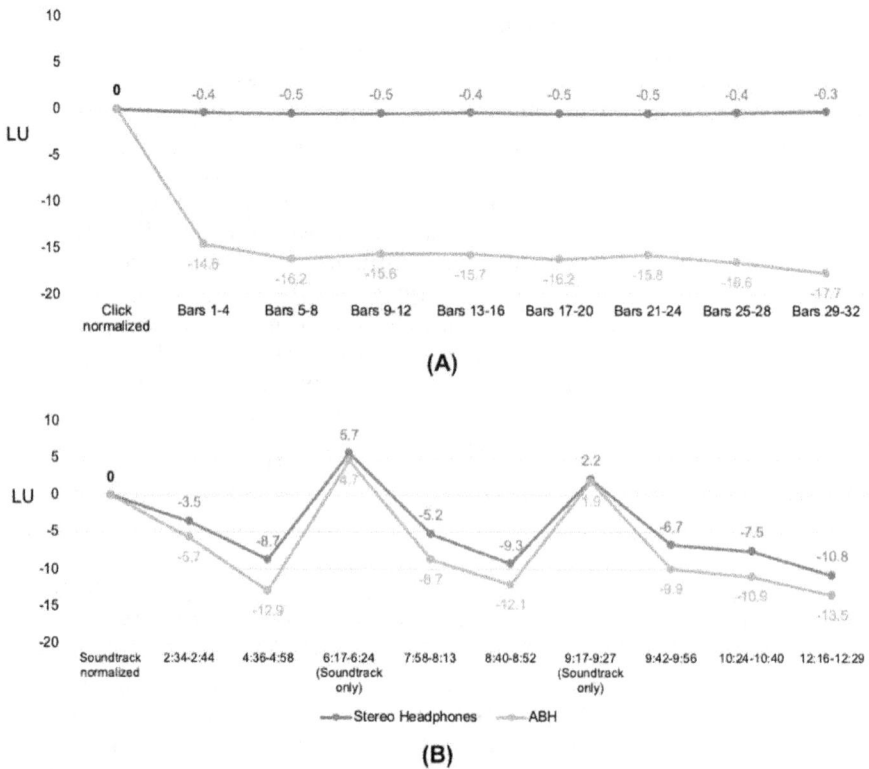

Figure 4.
Click-to-music loudness ratios (A) in the drummer's monitoring mix at key sections of the guitar and drums pop-rock duo, and (B) in the pianist's monitoring mix at key timings of Nicole Lizée's Hitchcock Études.

a click track on the left channel (B). We observe that for the chorus of the pop-rock duo, the drummer monitored the click track at 17.7 LU lower than the music with the ABH, versus at nearly the same loudness as the music at 0.3 LU with the stereo headphones. While the CMR decrease was less noticeable for the pianist's mix, we could observe that the ABH enabled a more dynamic headphone mix, and so a more expressive balance between the piano and soundtrack than the stereo headphones.

5. Discussion

5.1 What are the main challenges of using a monitoring wearable device while performing music?

Results from the questionnaires expand previous findings regarding the challenges that musicians face when performing with wearable monitoring devices [1]. In addition to being very sensitive to the sound quality of the headphone mix, performers also strongly value the technical quality and physical properties of the monitoring system. Moreover, results confirm that they develop strategies to cope with their discomfort. Indeed, only one out of the 21 respondents who answered the fourth question reported wearing headphones on both ears while performing. It should be noted that wearing only one earcup or half of both earcups is tiring for performers due to the asymmetry or layer of the auditory feedback. These findings thus reinforce the need to find monitoring solutions that overcome the challenges of traditional stereo headphones.

A large number of phrasings about negative musical consequences show that musicians are aware of the impact of poor monitoring setups on their performance. In this view, instrumentalists' comments during the case performance studies confirm that many do not expect to get a comfortable headphone mix in the studio [1] and that some of them come to the studio mentally prepared to face monitoring challenges. For instance, the drummer of the jazz trio explained that he usually expects to experience latency issues. However, while we know that monitoring mixes lead to different ways of performing, be the impact positive or negative [3], survey respondents surprisingly did not mention any positive musical consequences. Similarly, we noticed a reluctance from the participants who tested the BMR to detail the positive effects of their preferred monitoring condition on their musicality. These findings indicate that musicians and sound engineers should communicate more about monitoring systems to transcend the status quo. Also, results show that improvisers conceptualize their ideal monitoring system differently than orchestra conductors do, which corroborates with the need for engineers to adapt the design of monitoring systems as well as recording and live engineering sound choices [27] to the esthetic and culture of the performance context.

5.2 Could binaural technologies that are adapted to specific performance contexts enhance musicians' listening comfort, perception of a realistic auditory scene, and musical expression and creativity?

Across the seven comparison case studies, performers appreciated the listening comfort in binaural compared to stereo, and they expressed that the binaural rendering was more realistic than the stereo rendering. Nevertheless, in keeping with AMR and AAR definitions from the literature review [24, 26], the meaning of the realism concept varied depending on the augmentation type, from a convincing recreated spatial auditory scene in AMR to an auditory scene close to what performers heard in the real acoustics in AAR. This AAR realism definition was further researched by Soudoplatoff and Pras [7] who asked 15 sound engineers to describe how real they perceived the superposition of the binaural rendering of two soundscapes that were captured in the same room with the same microphone setup. The two soundscapes featured a jazz duo performance that was happening live in real time on the other side of the studio window, and a crowded ambiance that was recorded a few days before to give the illusion of a bar soundscape. Results showed that participants perceived "scene realism and a well-established illusion of being in a crowd." These outcomes call for future research that would assess the relevance of superposing a binauralized pre-recording of the synchronization cues in the venue to the music in performers' monitoring mixes.

For all AAR and AMR case performance studies, findings highlighted that the binaural conditions enabled all participants to be more collectively and cognitively involved in their creative tasks compared with the stereo conditions. This implies that our AAR setups could overcome social interaction challenges when wearing headphones and that an AMR setup could enhance social interaction among participants remotely located. Specifically, we suggest that participants were more *immersed* in performance [50] and that the free improvisers experienced a state of *flow* [63] since they performed longer takes with binaural monitoring. These research outcomes are important from an artistic perspective and should be made broadly available to musicians. Indeed, Menon noted from his studio experience as a rock guitarist that when controlling a personal monitoring mixer consumes more time than desired, musicians would rather cope with whatever they are hearing than to fix these issues. Therefore, we believe that greater learning around the impact of monitoring systems on the musicians' ability to be immersed in performance would motivate them to always ensure an optimized headphone mix.

In keeping with the findings of the BBC study [45], we found that the binaural rendering of the main array worked well to recreate the auditory scene of the film-scoring orchestra. In contrast, the binaural rendering of the same array was problematic in the jazz symphonic ensemble situation that featured complex interactions of room and instrument acoustics. Here we propose three solutions that could have helped reduce the drum leakage and better integrate the electric instruments and the double bass within the auditory scene. First, the percentage of natural reverberation in the mix could be manipulated by changing the balance between the main array and the spot microphones. Second, reverberation could be artificially re-created by using a real-time binaural room simulator with an object-based mixing device to wet the electric instruments and double bass, and thus enable their acoustic homogeneity with the rest of the auditory scene. Finally, using transparent glass panels could minimize the leakage of the drums while maintaining visual contact and giving all band members the illusion of playing in the same room. Moreover, the jazz symphonic ensemble situation reminded us that flaws in the quality and intercommunication setup of a monitoring system can be detrimental to the performance situation, for instance by increasing performers' stress [1], and/or by disempowering musicians while reinforcing the engineers' sound control [4]. This negative experience demonstrates that technological adaptation to the music performance context requires sound engineers to consider the overall studio context and researchers to ensure the success of the first trial.

5.3 Could binaural headphone monitoring systems increase the click-to-music ratio compared to stereo headphones?

With the ABH and BMR, we observed that performers required fewer synchronization cues in their headphone mix compared to the stereo conditions, due to enhanced binaural intelligibility and less sound masking effects [17]. The less dynamic nature of the stereo headphones brings musician to monitor a louder headphone mix overall, which is likely to damage their aural health over time [13]. Indeed, in addition to perceiving a more realistic spatial image, the extreme dynamic range differences between the click track and the music signal are enhanced by the binaural rendering, leading musicians to set their monitor level at a comfortable volume to enjoy the dips and valleys of the musical scenario. In this view, the drummer who used the ABH system got concerned about losing his hearing when he realized how loud his click track was in stereo. He mentioned to Menon that he would consider purchasing a wearable metronome to avoid using audible click tracks at such high volumes in the future. Similar findings appeared with the BMR system, as the jazz musicians and free improvisers could hear the music cues more distinctly without forcing in binaural compared to stereo. In particular, the jazz drummer and guitarist asked Bauer to increase the bass level in their monitoring mix when switching from binaural to stereo, as they explained that the bass was masked by other music elements in stereo. Also, whereas the world music performer removed one earcup in stereo while overdubbing, as the headphone mix content gradually got denser, he kept both earcups on throughout the recording process with the binaural condition. These findings also illustrate the challenge of controlling the balance of headphone mixes in stereo.

One of Soudoplatoff's motivations in developing a binaural monitoring solution was that a conductor from his professional network suffered from hyperacusis and tinnitus due to working with loud in-ear monitors when she toured with a symphonic orchestra mixed with electronic music in a dozen representations over a two-week period. The results of our seven comparison tests with three binaural monitoring setups encourage us to pursue this research to improve music performers' working conditions. One next step would be to focus on better integration of the click track in the monitoring mix by spatializing it to allow for its externalization with an appropriate localization distance. We believe that this advancement would enable musicians to monitor even less click track than with our BHT, BMR, ABH, and Copper and Martin [2]'s ATH, and thus to feel even more immersed in the performance. This should allow to reduce the cultural implications of the click track [14, 15]. Such an approach would thus treat the click track as AAR instead of AMR and would take full advantage of binaural unmasking capabilities.

6. Conclusion

Our research contributions that concur to show the potential of dynamic and static binaural monitoring solutions in enhancing performers' immersion in creative cognition are threefold. First, we constructed new theoretical knowledge on musicians' experiences when performing with headphones based on a multidisciplinary literature review and on survey responses from professional orchestra conductors and music improvisers. This knowledge provides acousticians with important insights to develop ecologically valid experimental protocols to assess innovative technologies in professional music performance contexts. Second, we designed one AMR and two AAR binaural monitoring solutions for which we detailed how we adapted their augmentation type, sound capture, and mixing approaches to distinct music performance contexts. Sound engineers can extend and modify these solutions to other contexts, within and beyond music performance. Third, we explored mixed-method approaches to assess our technologies in three different professional performance contexts through eight case studies. These approaches combined performers' feedback on their experience by comparing binaural versus stereo conditions, their choice of takes, and the measurement of click-to-music loudness ratios in both conditions. Discussed in terms of performers' listening comfort, their perception of a realistic auditory scene, and their musical expression and creativity, our case study outcomes could be integrated into close-ended feedback questionnaire in future studies that aim at assessing monitoring solutions based on performers' experience.

Because this series of studies drew out more insights into the positive influence of using binaural headphone monitoring for instrumentalists than for conductors, future comparisons between the two AAR solutions, namely BHT and ABH, will be pursued with professional conductors to determine which dynamic binaural solutions could best support their performance experience. Because our case study outcomes underline the positive influence of binaural monitoring over the music performance for instrumentalists but do not address the case of singers, further research will identify which binaural monitoring solution would best support professional singers' performance needs. Future research will also include tests with more musicians of different popular music genres to find solutions in terms of beat spatialization, which we know can be tricky in binaural, especially for hip hop [47].

To circumvent the hesitation to acknowledge the impact of monitoring technology on professional performers' musicality among practitioners, we encourage researchers to adopt a post-performance procedure, for instance through the analysis of performers' take choices a few weeks after the recording session to analyze the potential interconnection between the experimental condition and the best musical result [62]. This approach calls for conducting case studies that are fully integrated into real-life recording sessions that last several days. Also, future studies should further examine music performers' perception of *acoustic realism* when creating music in real-life situations with binaural monitoring, and how this perception depends on the cultural context (musical genre, ensemble's habits) and the acoustic situation (acoustic separation or not, size of the venue, amplified instruments or not). To that end, it would be interesting to compare three recording setups, such as instrumentalists being in the same room without headphone monitoring, instrumentalists being in the same room with binaural monitoring to augment their natural hearing, and instrumentalists being in separate rooms and hearing each other through binaural monitoring. This investigation could also help to refine the potential of AAR and AMR approaches from a practical point of view, and thus inform the design of future headphone monitoring systems. Furthermore, and with respect to the emerging concept of acoustic realism, a large longitudinal study should be conducted to identify the duration requirements of a training procedure with non-individualized HRTFs to reach an optimal level of performers' ability to externalize binaural audio cues, as well as to appreciate the intelligibility and comfort of a binaural mix. In that respect, using complex musical stimuli during the training procedure is advised, instead of non-ecological stimuli such as pink noise. At last, only few binaural music productions have been released on the market so far, and we hope that our research will inspire more sound engineers to explore binaural mixing techniques, and the music industry to give a chance to this 3D audio format.

Acknowledgements

We would like to thank all the questionnaire respondents and case study participants for their time, expertise, and useful insight into this research, as well as Dr. Terri Hron for her English review of our first submission. Also, we acknowledge the assistance and contribution of Dr. Georg Boenn and Chris Morris from the University of Lethbridge; Hervé Déjardin from Radio France; and Dr. Pascal Dietrich, Phil Kamp, Benedikt Krechel, Markus Pesch, and Dr. Roman Scharrer from KLANG: technologies to our technology designs and performance case studies. Menon's research assistantship and the publication processing charges for this chapter were funded by Pras' Partnership Development Grant of the Social Sciences and Humanities Research Council of Canada (SSHRC).

Author details

Valentin Bauer[1*], Dimitri Soudoplatoff[2], Leonard Menon[3] and Amandine Pras[4]

1 The Interdisciplinary Laboratory for Computer Sciences (LISN), Université Paris-Saclay, CNRS, France

2 Specialised Orchestral Conducting Master of Arts, Geneva University of Music (HEM), Switzerland

3 Schulich School of Music, McGill University, Canada

4 Department of Music, University of York, United Kingdom

*Address all correspondence to: valentin.bauer@lisn.fr

IntechOpen

References

[1] Williams A. I'm Not Hearing What You're Hearing: The Conflict and Connection of Headphone Mixes and Multiple Audioscapes. In: Frith S, Zagorski-Thomas S, editors. The Art of Record Production: An Introductory Reader for a New Academic Field. 1st ed. Farnham: Ashgate; 2012. p. 113-128

[2] Cooper A, Martin N. The impact of a prototype acoustically transparent headphone system on the recording studio performances of professional trumpet players. In: Hepworth-Sawyer R, Hodgson J, Paterson J, Toulson R, editors. Innovation in Music: Performance, Production, Technology, and Business. 1st ed. New York: Routledge; 2019. pp. 368-384. DOI: 10.4324/9781351016711-23

[3] Berg J, Johannesson T, Löfdahl M, Nykänen A. In-ear vs. loudspeaker monitoring for live sound and the effect on audio quality attributes and musical performance. In: Proceedings of the Audio Engineering Society Convention 142; 20-23 May 2017; Berlin. New York: Audio Engineering Society; 2017

[4] Brooks G, Pras A, Elafros A, Lockett M. Do we really want to keep the gate threshold that high? Journal of the Audio Engineering Society. 2021;69(4):238-260. DOI: 10.17743/jaes.2020.0074

[5] Berg J, Jullander S, Sundkvist P, Kjekshus H. The influence of room acoustics on musical performance and interpretation—A pilot study. In: Proceedings of the Audio Engineering Society Convention 140; 4-7 June 2016; Paris. New York: Audio Engineering Society; 2016

[6] Slaten WJ. Doing Sound: An Ethnography of Fidelity, Temporality, and Labor Among Live Sound Engineers. New York City: Columbia University; 2018

[7] Soudoplatoff D, Pras A. Augmented reality to improve orchestra conductors'

headphone monitoring. In: Proceedings of the Audio Engineering Society Convention 142; 20-23 May 2017; Berlin. New York: Audio Engineering Society; 2017

[8] Bauer V, Déjardin H, Pras A. Musicians' binaural headphone monitoring for studio recording. In: Proceedings of the Audio Engineering Society Convention 144; 23-26 May 2018; Milan. New York: Audio Engineering Society; 2018

[9] Menon L. Click-to-music ratio: Using active headphones to increase the gap. In: Proceedings of the Audio Engineering Society Convention 149; 27-30 October 2020; Online. New York: Audio Engineering Society; 2020

[10] Guastavino C, Katz BF, Polack J, Levitin DJ, Dubois D. Ecological validity of soundscape reproduction. Acta Acustica United With Acustica. 2004;91:333-341

[11] Donin N, Traube C. Tracking the creative process in music: New issues, new methods. Musicae Scientiae. 2016;20(3):283-286. DOI: 10.1177/1029864916656995

[12] Klapholz J. Fantasia: Innovations in sound. Journal of the Audio Engineering Society. 1991;39(1/2):66-70

[13] Chasin M. Musicians and the prevention of hearing loss. In: Proceedings of the 2018 Audio Engineering Society International Conference on Music Induced Hearing Disorders; 20-22 June 2018. Chicago Illinois. New York: Audio Engineering Society; 2018

[14] Cardassi L. In search of expressive time in mixed media works: Composer-performer collaboration, synchronization cues and customized click-tracks. In: Proceedings of the Encontros de Cognição Musical—Processos Criativos; 18-20 November

2020; Salvador: Universidade Federal da Bahia; 2020

[15] Cardassi L. Balancing musical and mechanical cues for synchronization in mixed media works and a model for customized click-tracks. ART Music Review.

[16] Blauert J. Spatial Hearing: The Psychophysics of Human Sound Localization. Revised ed. Cambridge: MIT Press; 1996. 502 p

[17] Oltheten W. Mixing with Impact: Learning to Make Musical Choices. 1st ed. New York: Routledge; 2018. 364 p. DOI: 10.4324/9781315113173

[18] Roessner S. The Beat Goes Static: A Tempo Analysis of US Billboard Hot 100# 1 Songs from 1955-2015. In: Proceedings of the Audio Engineering Society Convention 143; 18-20 October 2017; New York. New York: Audio Engineering Society; 2017

[19] Roginska A. Binaural audio through headphones. In: Roginska A, Geluso P, editors. Immersive Sound. 1st ed. New York: Routledge; 2017. pp. 88-123. DOI: 10.4324/9781315707525-5

[20] Lindeman RW, Noma H, De Barros PG. Hear-through and mic-through augmented reality: Using bone conduction to display spatialized audio. In: Proceedings of the 6th IEEE and ACM International Symposium on Mixed and Augmented Reality; 13-16 November 2007; Nara. Washington, DC: IEEE; 2007

[21] Fujise A. Investigation of practical compensation method for bone conduction headphones with a focus on spatialization. In: Proceedings of the 2018 Audio Engineering Society International Conference on Spatial Reproduction-Aesthetics and Science; 7-9 August 2018; Tokyo. New York: Audio Engineering Society; 2018

[22] May KR, Walker BN. The effects of distractor sounds presented through bone conduction headphones on the localization of critical environmental sounds. Applied Ergonomics. 2017;**61**: 144-158. DOI: 10.1016/j.apergo. 2017.01.009

[23] Albrecht R, Lokki T, Savioja L. A mobile augmented reality audio system with binaural microphones. In: Proceedings of Interacting with Sound Workshop: Exploring Context Aware, Local and Social Audio Applications; 30 August 2011; Stockholm. New Yok: ACM; 2011. pp. 7-11

[24] Härmä A, Jakka J, Tikander M, Karjalainen M, Lokki T, Hiipakka J, et al. Augmented reality audio for mobile and wearable appliances. Journal of the Audio Engineering Society. 2004;**52**(6):618-639

[25] Milgram P, Kishino F. A taxonomy of mixed reality visual displays. IEICE Transactions on Information and Systems. 1994;**77**:1321-1329

[26] McGill M, Brewster S, McGookin D, Wilson G. Acoustic transparency and the changing soundscape of auditory mixed reality. In: Proceedings of the 2020 CHI Conference on Human Factors in Computing Systems; 25-30 April 2020; Honolulu. New-York: ACM; 2020. pp. 1-16

[27] Pras A, Guastavino C, Lavoie M. The impact of technological advances on recording studio practices. Journal of the American Society for Information Science and Technology. 2013;**64**(3):612-626. DOI: 10.1002/asi.22840

[28] König FM. New measurements and psychoacoustic investigations on a headphone for TAX/HDTV/Dolby-surround reproduction of sound. In: Proceedings of the Audio Engineering Society Convention 98; 25-28 February 1995; Paris. New York: Audio Engineering Society; 1995

[29] Møller H. Fundamentals of binaural technology. Applied Acoustics. 1992;**36**(3-4):171-218

[30] Best V, Baumgartner R, Lavandier M, Majdak P, Kopčo N. Sound externalization: A review of recent research. Trends in Hearing. 2020;**24**: 1-14. DOI: 10.1177/2331216520948390

[31] Mendonça C, Campos G, Dias P, Vieira J, Ferreira JP, Santos JA. On the improvement of localization accuracy with non-individualized HRTF-based sounds. Journal of the Audio Engineering Society. 2012;**60**:821-830

[32] Chatzidimitris T, Gavalas D, Michael D. SoundPacman: Audio augmented reality in location-based games. In: Proceedings of the 18th Mediterranean Electrotechnical Conference; 18-20 April 2016; Limassol. New York: IEEE; 2016. pp. 1-6

[33] Katz BF, Kammoun S, Parseihian G, Gutierrez O, Brilhault A, Auvray M, et al. NAVIG: Augmented reality guidance system for the visually impaired. Virtual Reality. 2012;**16**:253-269. DOI: 10.1007/s10055-012-0213-6

[34] Naphtali D, Rodkin R. Audio augmented reality for interactive soundwalks, sound art and music delivery. In: Filimowicz M, editor. Foundations in Sound Design for Interactive Media. 1st ed. New York: Routledge; 2019. pp. 300-332. DOI: 10.4324/9781315106342

[35] Mariette N. Human factors research in audio augmented reality. In: Huang W, Alem L, Livingston MA, editors. Human Factors in Augmented Reality Environments. 1st ed. New York: Springer; 2013. pp. 11-32. DOI: 10.1007/978-1-4614-4205-9

[36] Nagele AN, Bauer V, Healey PG, Reiss JD, Cooke H, Cowlishaw T, et al. Interactive audio augmented reality in participatory performance. Frontiers in Virtual Reality. 2020;**1**:46. DOI: 10.3389/frvir.2020.610320

[37] Moustakas N, Floros A, Grigoriou N. Interactive audio realities:

An augmented/mixed reality audio game prototype. In: Proceedings of the Audio Engineering Society Convention 130; 13-16 May 2011; London. New York: Audio Engineering Society; 2011

[38] Zimmermann A, Lorenz A. LISTEN: A user-adaptive audio-augmented museum guide. User Modeling and User-Adapted Interaction. 2008;**18**: 389-416

[39] Mariette N, Katz BF, Boussetta K, Guillerminet O. SoundDelta: A study of audio augmented reality using WiFi-distributed Ambisonic cell rendering. In: Proceedings of the Audio Engineering Society Convention 128; 22-25 May 2010; London. New York: Audio Engineering Society; 2010

[40] Goony A. Les HRTF appliquées au retour de scene par "in-ear monitors" [master thesis]. Saint Denis: École nationale supérieure Louis-Lumière; 2010

[41] Zea E. Binaural In-Ear Monitoring of acoustic instruments in live music performance. In: Proceedings of the 15th International Conference on Digital Audio Effects; 30 November – 3 December 2012; Trondheim. Trondheim: Norwegian University of Science and Technology; 2012. pp. 1-8.

[42] Field A. Hearing the past in the present: An augmented reality approach to site reconstruction through architecturally informed new music. In: Schofield J, Maloney L, editors. Music and Heritage. 1st ed. London: Routledge; 2021. pp. 212-221

[43] Murphy D, Shelley S, Foteinou A, Brereton J, Daffern H. Acoustic Heritage and Audio Creativity: the Creative Application of Sound in the Representation, Understanding and Experience of Past Environments. Internet Archaeology. 2017;**44**. DOI: 10.11141/ia.44.12

[44] Gebhardt M, Kuhn C, Pellegrini R. Headphones Technology for Surround

Sound Monitoring–A Virtual 5.1 Listening Room. In: Proceedings of the Audio Engineering Society Convention 122; 5-8 May 2007; Vienna. New York: Audio Engineering Society; 2007

[45] Parnell T, Pike C. An efficient method for producing binaural mixes of classical music from a primary stereo mix. In: Proceedings of the Audio Engineering Society Convention 144; 23-26 May 2018; Milan. New York: Audio Engineering Society; 2018

[46] Walton T. The overall listening experience of binaural audio. In: Proceedings of the 4th International Conference on Spatial Audio; 7-10 September 2017; Graz. VDT&IEM; 2017. pp. 170-177

[47] Turner K, Pras A. Is Binaural Spatialization the Future of Hip-Hop?. In: Proceedings of the Audio Engineering Society Convention 147; 21-24 October 2019; New York. New York: Audio Engineering Society; 2019

[48] Corbin J, Strauss A. Strategies for qualitative data analysis. In: Corbin J, Strauss A, editors. Basics of Qualitative Research. Techniques and Procedures for Developing Grounded Theory. 3rd ed. New York: SAGE Publications Inc; 2008. 434 p. DOI: 10.4135/9781452230153.n4

[49] Pras A, Guastavino C. The role of music producers and sound engineers in the current recording context, as perceived by young professionals. Musicae Scientiae. 2011;15(1):73-95. DOI: 10.1177/1029864910393407

[50] Agrawal S, Simon A, Bech S, Bærentsen K, Forchhammer S. Defining immersion: Literature review and implications for research on immersive audiovisual experiences. Journal of Audio Engineering Society. 2019;68(6):404-417. DOI: 10.17743/jaes.2020.0039

[51] Cummings JJ, Bailenson JN, Fidler MJ. How immersive is enough? A foundation for a meta-analysis of the effect of immersive technology on measured presence. Media Psychology. 2016;19:272-309. DOI: 10.1080/15213269.2015.1015740

[52] Heeter C. Being there: The subjective experience of presence. Presence: Teleoperators & Virtual Environments. 1992;1(2):262-271

[53] Messonnier JC, Lyzwa JM, Devallez D, de Boishéraud C. Object-based audio recording methods. In: Proceedings of the Audio Engineering Society Convention 140; 4-7 June 2016; Paris. New York: Audio Engineering Society; 2016

[54] Baskind A, Messonnier JC, Lyzwa JM. Bipan: An experimental mixing tool for 3D-audio on headphones with open-source head tracker. In: Proceedings of the 29th Tonmeistertagung-VDT International Convention; 17-20 November 2016; Cologne. Köln: Verband Deutscher Tonmeister e.V.; 2016

[55] Warusfel O. LISTEN HRTF database. Available from: http://recherche.ircam.fr/equipes/salles/listen/ [Accessed: July 11, 2021]

[56] Hendrickx E, Stitt P, Messonnier JC, Lyzwa JM, Katz B, De Boishéraud C. Influence of head tracking on the externalization of speech stimuli for non-individualized binaural synthesis. The Journal of the Acoustical Society of America. 2017;141:2011-2023. DOI: 10.1121/1.4978612

[57] Nicol R, Gros L, Colomes C, Roncière E, Messonnier JC. Etude comparative du rendu de différentes techniques de prise de son spatialisée après binauralisation. In : Proceedings of the 13ème Congrès Français d'Acoustique; 11-15 April 2016; Le Mans. Paris: Société Française d'Acoustique; 2016. pp. 211-217

[58] Lester M, Boley J. The effects of latency on live sound monitoring.

In: Proceedings of the Audio
Engineering Society Convention 123;
5-8 October 2007; New York. New York:
Audio Engineering Society; 2007

[59] Stitt P, Hendrickx E, Messonnier JC,
Katz B. The influence of head tracking
latency on binaural rendering in
simple and complex sound scenes. In:
Proceedings of the Audio Engineering
Society Convention 140; 4-7 June 2016;
Paris. New York: Audio Engineering
Society; 2016

[60] Brungart D, Kordik AJ, Simpson BD.
Effects of headtracker latency in virtual
audio displays. The Journal of the
Acoustical Society. 2006;**54**(1/2):32-44

[61] Eley N. Classification of HRTFs
using perceptually meaningful
frequency arrays. In: Proceedings of the
Audio Engineering Society Convention
147; 21-24 October 2019; New York. New
York: Audio Engineering Society; 2019

[62] Pras A, Guastavino C. The impact of
producers' comments and musicians'
self-evaluation on perceived recording
quality. Journal of Music, Technology &
Education. 2013;**6**(1):81-101. DOI:
10.1386/jmte.6.1.81_1

[63] Csikszentmihalyi M. Flow: The
Psychology of Optimal Experience. New
York: Harper & Row; 1990. p. 303

www.ingramcontent.com/pod-product-compliance
Lightning Source LLC
Chambersburg PA
CBHW081538190326
41458CB00015B/5580